I0487063

A Curative
Cancer Treatment

By

Isaac Lasley

A Curative Cancer Treatment

Isaac Lasley

Lulu.com Edition

This book was written to share my curative cancer treatment hypotheses with the world to hopefully spark a worldwide curative cancer treatment resource that will help to enable humanity to end its pain caused as a result of this terrible disease of cancer. The majority of the online informational resources utilized throughout this book came from The National Center for Biotechnology Information and Cancer GeneticsWeb. The pictures are from noted sources that have been publicized online.

Table of Contents

Introduction

I am a research scientist who studies a broad range of subjects. I wrote this book to share my curative cancer treatment with the world as it has such great potential. I must begin by noting my appreciation for my father, David Lasley, for lighting the fire under my ass that helped me to ever attempt to create a curative cancer treatment process. I also thank Dr. Jim Bonacum, a professor of microbiology at the University of Illinois Springfield, for inspiring me to write this book in order to copyright and share my ideas with the world rather than enter my hypotheses into an inventor's contest. Jim was my inspiration for this book and before meeting him on that summer day in 2007 to propose my hypotheses for approval to be entered into the inventor's contest I had never seen a man's eyes bug out of his face far enough to push his glasses down his nose.

What Is Cancer and Why Does It Occur?

The National Cancer Institute describes cancer as a term used for disease in which abnormal cells divide without control and are able to invade other tissues. Cancer cells can spread to other parts of the body through the blood and lymph systems. There are more than 100 types of cancer most named for its organ/cell type. Broader types of cancer include carcinoma, sarcoma, leukemia, lymphoma, myeloma, and CNS cancers.

Carcinoma occurs in the skin or tissues that line or cover internal organs.

Sarcoma occurs in bone, cartilage, fat, muscle, blood vessels, or other connective or supportive tissue.

Leukemia occurs in blood-forming tissue such as the bone marrow and causes large numbers of abnormal blood cells to be produced and enter the blood.

Lymphoma and myeloma occurs in the cells of the immune system.

Central nervous system cancers occur in the tissues of the brain and spinal cord [1].

The American Cancer Society notes that cancer is a collection of more than 100 diseases in which cells in part of the body begin to grow out of control. Many types of cancer begin as a result of abnormal cells growing out of control. Untreated cancers can cause serious illness or death [2].

The American Society of Clinical Oncology describes cancer as more than 100 diseases characterized by uncontrolled abnormal growth of cells. These cells form a lump or mass called a tumor. Some cancers, like blood cancers, don't form tumors. Tumors can be benign or malignant. Benign tumors may grow, but they do not spread to other parts of the body and are usually not life threatening. Malignant tumors grow and invade other tissues in the body.

Sometimes cancer will spread to the lymph nodes. A lymph node is a tiny, bean-shaped organ that filters the flow of lymph, the clear fluid that plays a role in the body's immune system. Lymph nodes are located in clusters in different parts of the body, such as the neck, groin area, and under the arms. Cells from malignant tumors can also break away and travel to other parts of the body, where they can continue to grow. This process is called metastasis. Metastatic cancer is named for the part of the body where it started. For example, if breast cancer spreads to the lungs, it is called metastatic breast cancer, not lung cancer. Cancer can begin almost anywhere in the body. Tumors are named for the type of cell where the cancer started. As, carcinomas begin in the skin or tissue that covers the surface of internal organs and glands. Sarcomas begin in the connective tissue, such as muscle, fat, cartilage, or bone [3].

I understand through my research that cancer is formed out of what were once normal human cells that have mutated in some way to become abnormal, then mutating even more to become cancerous in nature. The cancerous cells may be benign or malignant, meaning that the cells are mobile and have the ability to spread to other areas of the body. The primary causes of cancer include the environment in which one works or resides, personal habits like smoking, and genetic variations. Cancer cells were at one point normally functioning cells of the human body before mutating to form abnormal, cancerous cells that then reproduce as cancer.

What Is The Core Cause Of Cancers?

Cancer occurs after a normal cell changes and mutates in some way to begin functioning abnormally, then reproducing abnormal cells. Mutations that arise in normal cell DNA can cause cells to change becoming abnormal cancerous cells, which still have identifying DNA, but function improperly, reproducing cancerous cells with abnormal DNA. Of course the DNA from various cells throughout the body may begin to mutate for a variety of reasons, such as those causal factors noted for each cancer in this chapter. By understanding all of the risk factors for each cancer type we are better able to understand how to help prevent cancers from occurring and how to more effectively treat cancers when they do arise. To cure cancer science must note the core cause of all cancer types as it forms as a result of genetic mutation followed by reproduction of the mutated or genetically altered human cells [cancer cells]. Considering that cancer cells began from once normal human cells we can better understand why the human immune system usually does not recognize or react to cancer cells as a threat to homeostasis.

What Cancer Treatments Are Currently Used?

Common cancer treatments currently used; such as chemotherapy, radiation, and surgery; are on some level harmful to the healthy cells of the body, as well as the cancerous cells. They all stop cancer cells at some level from multiplying by using drugs, radiation, surgery, stem cell/bone marrow transplant, photodynamic therapy, hyperthermia, cryotherapy, angiogenesis inhibitory therapy, and/or a gene therapy for kidney cancer. None of these treatments can fully guarantee that the cancer cells will stop dividing and multiplying, although some chemotherapy drugs and radiation do slow or help to stop cancer cell reproduction, while surgery tries to remove all of the cancer cells from the body. Photodynamic therapy uses light sensitizers. Of course stem cell/bone marrow transplant may also be used to replace cancerous blood forming tissue. Many negative side effects result from common cancer treatments which in turn result in the need for complementary and alternative treatments to help cancer victims to better deal with such side effects.

Some of the less common cancer treatments available have the potential to be more effective, with fewer side effects than the most commonly known treatments. I hope that research is continued on those treatments to continue to improve them. I believe that gene therapy has much greater potential, than has been realized, as I have noted throughout this chapter. We live in the modern age when an updated human genome is available on the internet, DNA Modification is possible, and genetic cloning has occurred. I still remember when I was in high school and Dolly the sheep was the first successfully cloned mammal. Of course I am getting off subject a bit, but cloning will likely be important once the curative cancer cocktails are brewed up. First, we must review our understanding of the problem in front of us before we can solve it.

I believe that cancer may need to be treated on a genetic level. While there may be hundreds of various specific types of cancers, there are literally thousands of genetic types of cancers. For example, while maybe around two out of a hundred people diagnosed with lung cancer may have cancers with the same identifying genetic code, that fact can be considered an advantage rather than a disadvantage. We now may note that two people can be diagnosed with the same genetic type of cancer. This knowledge can be noted as vitally useful if properly understood. That means that different people can experience the same cell mutations and have the same genetic types of cancer.

Since cancer cells were initially normal human cells before they mutated and began to reproduce as cancer cells we can conclude that these cells were initially able to be removed from the body via the cellular process of apoptosis before becoming cancerous. Apoptosis is the natural death of the cells in which the remaining cellular material is destroyed and removed from the body after cell death. Necrosis is another form of cell death, which results in an inflammatory response as the dead cellular material is not removed from the body or destroyed within the body and as a result builds up within tissue. Of course now that these cancer cells have mutated, in other words changed their genetic composition [DNA], they are now no longer capable of apoptosis. Or are they? If we can genetically modify the DNA in cancerous cells in the proper manner we can likely induce apoptosis!! How is that type of genetic modification accomplished? Every human being on the planet has a genetic map contained within the 23 chromosome pairs that can be found in each cell of the human body. Every individual received 23 chromosomes from their mother and their father, as only one chromosome may be unique to each individual which is the X or Y sex chromosome that was received from the father that pairs up with the mother's X sex chromosome. With this understanding it can be noted that all humanity has 45 of 46 chromosomes or 22 of 23 chromosome pairs in common, which can be noted as a very advantageous fact. Each chromosome is filled with genes that are nothing more than defined strands of deoxyribonucleic acid [DNA] that contain the instructions to make a specific protein or closely related group of proteins, which can be noted as inactive or activated [4]. As the human body is initially created from two strands of 23 chromosomes, one strand from a mother and one from a father, that simply divided and multiplied again and again to create a human body. As a result every cell in the human body has the same DNA and the same genes available within each.

The process of apoptosis involves the stimulation of certain cellular genes that initiate the release of cytochrome C and certain caspases from the cell mitochondria. Intrinsic apoptosis can occur from within the cell, as the p53 gene signals the process to begin as a result of stressors on the cell or the cell has simply divided and multiplied as many times as it can. This is the natural or most common way that cell suicide via apoptosis occurs as the BAX and/or BAK genes are activated and stimulate the mitochondrial release of caspases. Extrinsic apoptosis can occur independently of the p53 gene as a result of stimulation of the cellular death receptors by coordinating death receptor ligands to begin the process as caspases are released. Below are diagrams of both intrinsic and extrinsic apoptosis pathways in the cell.

Intrinsic Apoptosis

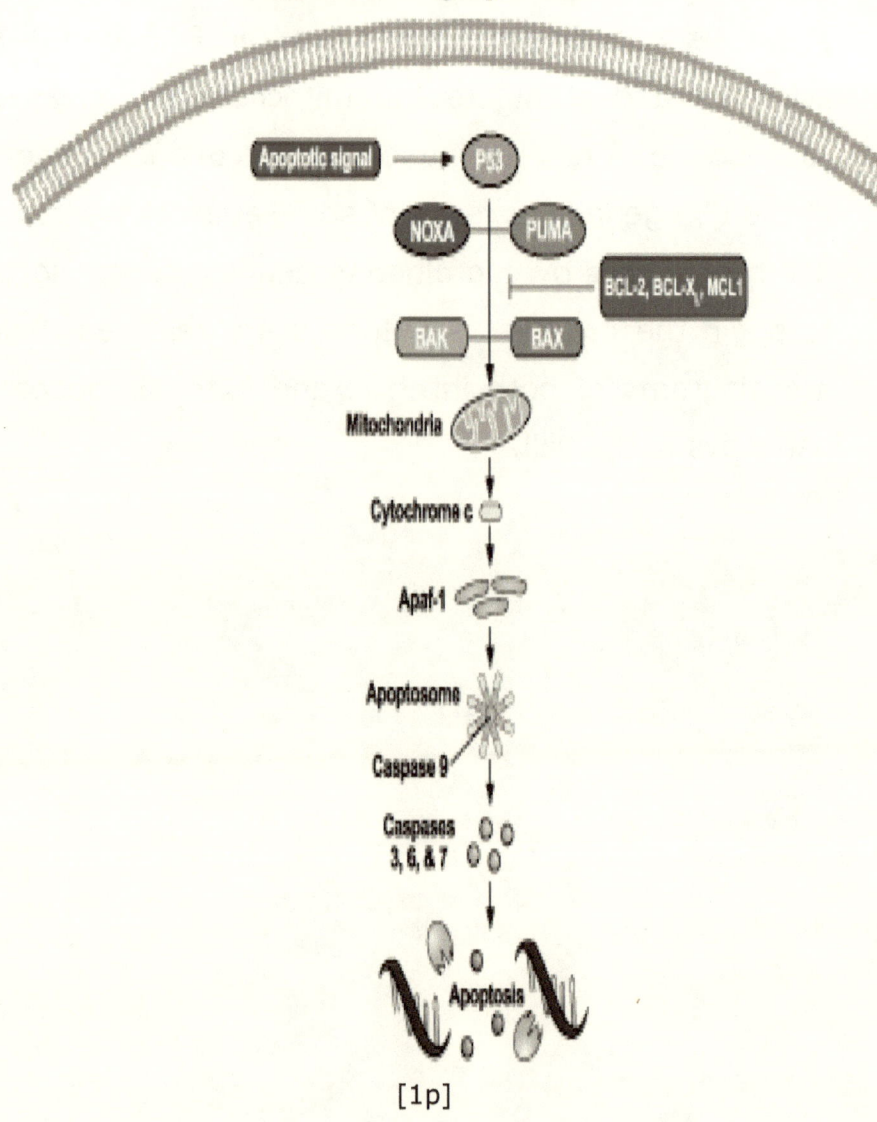

[1p]

Extrinsic Apoptosis

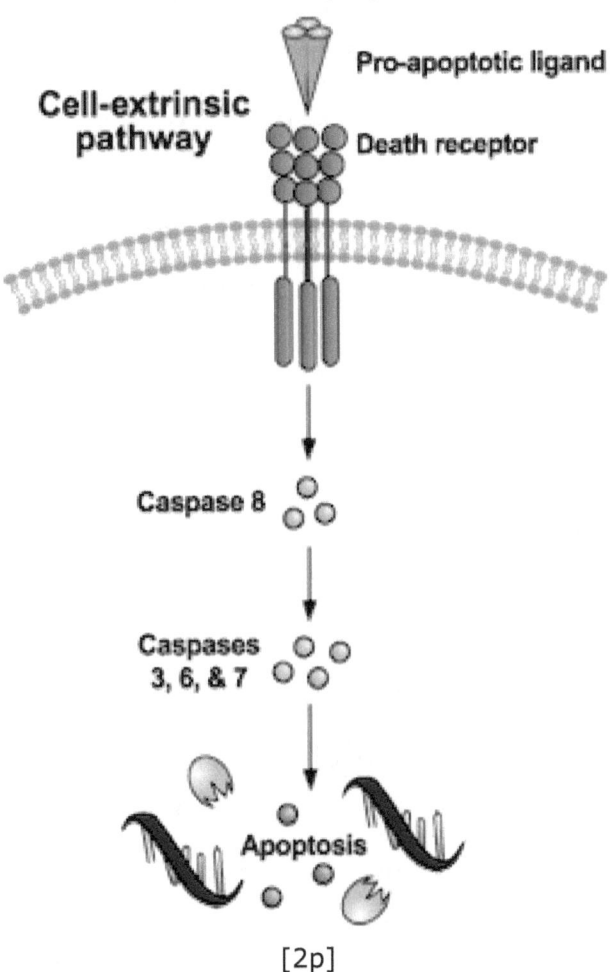

[2p]

What Genes Can Be
Activated To Stimulate Apoptosis?

Below I have noted genes that are noted by the "Unigene Project" to be apoptosis inhibitors, apoptosis regulators, apoptosis modulators, apoptosis facilitators, apoptosis inducers, apoptosis receptor ligands, apoptosis receptors, and apoptosis associates. I have listed the apoptosis genes and noted the chromosomes on which they exist. I then noted each type of cancer that I have reviewed, listing the apoptosis genes that are normally located within the healthy cell types of each before the cells mutated to become cancerous. After that I have noted the genes that are commonly mutated for each type of cancer reviewed. I will then review how my curative cancer treatment hypothesis may incorporate many of the noted apoptotic genes.

The mitochondrion within the cells is the key organelle that helps intrinsic apoptosis to occur. The proper mitochondria stimulus is the key to apoptosis in cancer cells. Once the mitochondrion is stimulated properly to release cytochrome c and the proper caspases [CASP] apoptosis has begun within the cell. As apoptosis progresses a cell and all of its materials are naturally removed from the body.

Extrinsic apoptosis occurs as apoptosis receptor ligands stimulate coordinating apoptosis receptors within each cell. Once the apoptosis receptor in the cell is properly stimulated caspases are released and apoptosis occurs as the cell is destroyed and flushed from the body. This process of apoptosis is much simpler than the process of intrinsic apoptosis. Although, it may or may not be as difficult to initiate, but that will be noted during experiments as both should be tested to prove which is more advantageous.

Each of the various types of apoptosis genes are noted for function. Apoptosis inhibitors prevent apoptosis from occurring. Apoptosis regulators are essential to regulate whether apoptosis occurs or does not occur. Apoptosis modulators are a necessary part of the process of apoptosis. Apoptosis facilitators mediate cell death as it occurs. Apoptosis inducers promote apoptosis to occur. Apoptosis receptors exist on the exterior of a cell and can be stimulated to begin the cell death process by coordinating apoptosis receptor ligands. Apoptosis associates are involved in the process of apoptosis after it has begun. Below are many noted apoptotic genes that may be stimulated or inhibited in an effort to promote apoptosis in cancer cells and their chromosome location.

Apoptosis Inhibitory Genes

API1 - 11q22 [5, 6]

API2 - 11q22 [7, 8]

API3 - Xq25 [9, 10]

API4 - 17q25 [11, 12]

API5 -11p11.2 [13, 14]

AVEN - 15q13.1 [15, 16]

BCL2 - 18q21.3 [17, 18]

BCL2A1 - 15q24.3 [19, 20]

BCLX - 20q11.22 [21, 22]

BIRC6 - 2p22-p21 [23, 24]

BIRC7 - 20q13.3 [25, 26]

CIAPIN1 - 16q13-q21 [27, 28]

FAIM - 3q22.3 [29]

FAIM2 - 12q13 [30, 31]

FAIM3 - 1q32.1 [32]

MCL1 - 1q21 [33, 34]

NAIP - 5q13.1 [35, 36]

PPP1R13L - 19q13.32 [37, 38]

PPP1R15A - 19q13.2 [39, 40]

SYVN1 - 11q13 [41, 42]

TRIAP1 - 12q24.31 [43]

Apoptosis Regulator Genes

BCLG - 12p12 [44]

BCLW - 14q11.2-q12 [45, 46]

BFAR - 6p13.12 [47]

CCAR1 - 10q21.3 [48, 49]

CFLAR - 2q33-q34 [50, 51]

PAWR - 12q21 [52, 53]

p53 - 17p13.1 [54, 55]

Apoptosis Modulator Genes

MOAP1 - 14q32 [56, 57]

PUMA - 19q13.3-q13.4 [58, 59]

Apoptosis Facilitator Genes

BCL2L10 - 15q21 [60, 61]

BCL2L11 - 2q13 [62, 63]

BCL2L13 - 22q11 [64]

BCL2L14 -12p13-p12 [65]

Apoptosis Inducer Genes

AIFM1 - Xq25-q26 [66, 67]

AIFM2 - 10q22.1 [68, 69]

APITD1 - 1p36.22 [70, 71]

BAD - 11q13.1 [72, 73]

BAK - 6p21.3 [74, 75]

BAX - 19q13.3-13.4 [76, 77]

BCL2L15 - 1p13.2 [78]

BID - 22q11.1 [79, 80]

BIK - 22q13.31 [81, 82]

BIM - 2q12-q13 [83, 84]

BMF - 15q24 [85, 86]

BOK - 2q37.3 [87, 88]

CIDEB - 14q12 [89, 90]

HRK - 12q24.22 [91, 92]

MRPS30 - 5p12-q11 [93, 94]

NAIF1 - 9q34.11 [95, 96]

NOXA - 18q21.32 [97, 98]

PDCD1 - 2q37.3 [99, 100]

PDCD2 - 6q27 [101, 102]

PDCD4 - 10q24 [103, 104]

PDCD5 - 19q12-q13.1 [105, 106]

PDCD6 - 5p15.33 [107, 108]

PDCD6IP - 3p22.3 [109, 110]

PDCD7 - 15q22.31 [111, 112]

PDCD10 - 3q26.1 [113, 114]

PDCD11 - 10q24.2-q25.1 [115, 116]

PERP - 6q24 [117, 118]

SIVA1 - 14q32.33 [119, 120]

TP53AIP1 - 11q24 [121, 122]

Apoptosis Receptor Ligand Genes

FASLG – 1q23 [DR1 [FAS] Ligand, TNFSF6, CD95L] [123, 124]

Lymphotoxin Alpha – 6p21.3 [DR2 Ligand, TNFSF1, TNFB, TNF-BETA] [125, 126]

APO3L – 17p13.1, 17p13.3 [DR3 Ligand, TNFSF12, TWEAK] [127, 128]

APO2L – 3q26 [DR4 and DR5 Ligand, TNFSF10, TRAIL] [129, 130]

PDCD1LG1 – 9p24 [Programmed Cell Death 1 Ligand 1] [CD274] [131, 132]

PDCD1LG2 – 9p24.2 [Programmed Cell Death 1 Ligand 2] [133, 134]

Apoptosis Receptor Genes

DR1 – 10q24.1 [Death Receptor 1] [135, 136]

DR2 – 12p13.2 [137, 138]

DR3 – 1p36.3 [139, 140]

DR4 – 8p21 [141, 142]

DR5 – 8p22-p21 [143, 144]

Apoptosis Associate Genes

AATF – 17q11.2-q12 [145, 146]

AATK – 17q25.3 [147, 148]

AEN – 15q26.1 [149, 150]

CASP1– 11q23 [151, 152]

CASP2 – 7q34-q35 [153, 154]

CASP3 – 4q34 [155, 156]

CASP4 – 11q22.2-q22.3 [157, 158]

CASP5 – 11q22.2-q22.3 [159, 160]

CASP6 – 4q25 [161, 162]

CASP7 – 10q25 [163, 164]

CASP8 – 2q33-q34 [165, 166]

CASP9 – 1p36.3-p36.1 [167, 168]

CASP10 – 2q33-q34 [Apoptosis-Related Cysteine Peptidase] [169, 170]

CASP14 – 19p13.1 [Apoptosis-Related Cysteine Peptidase] [171, 172]

DBNDD2 – 20q13.12, 20q12 [SCF Apoptosis Response Protein 1] [173, 174]

FADD – 11q13.3 [MORT1] [FAS-Associated Protein with Death Domain] [175, 176]

PIDD – 11p15.5 [LRDD] [p53 Induced Protein with Death Domain] [177, 178]

THAP1 – 8p11.21 [THAP Domain Containing Apoptosis Associate Protein 1] [179, 180]

THAP2 – 12q21.1 [THAP Domain Containing Apoptosis Associate Protein 2] [181, 182]

THAP3 – 1p36.31 [THAP Domain Containing Apoptosis Associate Protein 3] [183, 184]

TRADD – 16q22 [TNFR1 Associated Death Domain] [185, 186]

Which Apoptosis Genes
Are On Each Chromosome?

This should be noted in order to realize if there is a single chromosome or close group of chromosomes that contain the proper genes that may be modified in an effort to stimulate apoptosis in cancer cells. Cancer can likely be cured by effectively stimulating the apoptosis genes that are still available within the cancer cell chromosomes.

Chromosome 1

Apoptosis Inhibitors – FAIM3, MCL1

Apoptosis Inducers – APITD1, BCL2L15

Apoptosis Receptor Ligands – FASLG

Apoptosis Receptors – DR3

Apoptosis Associates – CASP9, THAP3

Chromosome 2

Apoptosis Inhibitors – BIRC6

Apoptosis Regulators – CFLAR

Apoptosis Facilitators – BCL2L11

Apoptosis Inducers – BIM, BOK, PDCD1

Apoptosis Associates – CASP8, CASP10

Chromosome 3

Apoptosis Inhibitors – FAIM

Apoptosis Inducers – PDCD6IP, PDCD10

Apoptosis Receptor Ligands – APO2L

Chromosome 4

Apoptosis Associates – CASP3, CASP6

Chromosome 5

Apoptosis Inhibitors – NAIP

Apoptosis Inducers – MRP30, PDCD6

Chromosome 6

Apoptosis Regulators – BFAR

Apoptosis Inducers – BAK, PDCD2, PERP

Apoptosis Receptor Ligands – Lymphotoxin Alpha

Chromosome 7

Apoptosis Associates – CASP2

Chromosome 8

Apoptosis Receptors – DR4, DR5

Apoptosis Associates – THAP1

Chromosome 9

Apoptosis Inducers – NAIF1

Apoptosis Receptor Ligands – PDCD1LG1, PDCD1LG2

Chromosome 10

Apoptosis Regulators – CCAR1

Apoptosis Inducers – AIFM2, PDCD4, PDCD11

Apoptosis Receptors – DR1

Apoptosis Associates – CASP7

Chromosome 11

Apoptosis Inhibitors – API1, API2, API5, SYVN1

Apoptosis Inducers – BAD, TP53AIP1

Apoptosis Associates – CASP1, CASP4, CASP5, FADD, PIDD

Chromosome 12

Apoptosis Inhibitors – FAIM2, TRIAP1

Apoptosis Regulators – BCLG, PAWR

Apoptosis Facilitators – BCL2L14

Apoptosis Inducers – HRK

Apoptosis Receptors – DR2

Apoptosis Associates – THAP2

Chromosome 13

No Apoptosis Genes Noted

Chromosome 14

Apoptosis Regulators – BCLW

Apoptosis Modulators – MOAP1

Apoptosis Inducers – CIDEB, SIVA1

Chromosome 15

Apoptosis Inhibitors – AVEN, BCL2A1

Apoptosis Facilitators – BCL2L10

Apoptosis Inducers – BMF, PDCD7

Apoptosis Associates – AEN

Chromosome 16

Apoptosis Inhibitors – CIAPIN1

Apoptosis Associates – TRADD

Chromosome 17

Apoptosis Inhibitors – API4

Apoptosis Regulators – TP53

Apoptosis Receptor Ligands – APO3L

Apoptosis Associates – AATF, AATK

Chromosome 18

Apoptosis Inhibitors – BCL2

Apoptosis Inducers – NOXA

Chromosome 19

Apoptosis Inhibitors – PPP1R13L, PPP1R15A

Apoptosis Modulators – PUMA

Apoptosis Inducers – BAX, PDCD5

Apoptosis Associates – CASP14

Chromosome 20

Apoptosis Inhibitors – BCLX, BIRC7

Apoptosis Associates – DBNDD2

Chromosome 21

No Apoptosis Genes Noted

Chromosome 22

Apoptosis Facilitators – BCL2L13

Apoptosis Inducers – BID, BIK

Chromosome X

Apoptosis Inhibitors – API3

Apoptosis Inducers – AIFM1

Which Apoptosis Genes
Are Mutated In Each Cancer Type?

The following are the specific genes that commonly mutate within cells and organs to begin cancerous growth. We should check these genes to note the apoptosis genes that are mutated within the various types of cancer cells that were once normal human tissue/cells. All of the noted apoptosis genes that may have mutated within each specific cancer type are bold and underlined. These cancer genes are all noted on "Cancer Genetics Web".

Bladder Cancer – P53 [187]

Osteosarcoma Bone Cancer – P53 [188]

Brain/CNS Cancer – P53 [189]

Breast Cancer – P53, API4 [190]

Colorectal Cancer – DR5, P53, BCL2, BAX [191]

Kidney Cancer – None [192]

Acute Lymphocytic Leukemia – DR1, BAX [193]

Acute Myeloid Leukemia – BCL2, BAX [194]

Chronic Lymphocytic Leukemia – P53, BAX [195]

Chronic Myeloid Leukemia – None [196]

Eosinophilic Leukemia – P53 [197]

Liver Cancer – P53, API4 [198]

Lung Cancer – P53, API4, BCL2 [199]

Hodgkin Lymphoma – BCL2 [200]

Non-Hodgkin Lymphoma – P53, API4, BCL2 [201]

Multiple Myeloma – None [202]

Ovarian Cancer – P53, BCL2 [203]

Pancreatic Cancer – BAX [204]

Prostate Cancer – P53 [205]

Melanoma Skin Cancer – P53 [206]

Basal Cell Carcinoma Skin Cancer – None [207]

Stomach Cancer – P53, BAX [208]

Testicular Cancer – None [209]

Thyroid Cancer – None [210]

Uterine Cancer – P53 [211]

Which Apoptosis Genes
Are Commonly Mutated In Cancer?

The BAX and p53 are the intrinsic apoptosis genes that are commonly mutated in cancer cells. Apoptosis can be stimulated through alternative pathways other than through the stimulation of p53. That is the key to curing cancer, initiating apoptosis in cancer cells through the activation or inhibition of apoptotic genes other than p53, as it is commonly mutated. It should be noted that p53 and BAX are both mutated in only chronic lymphocytic leukemia and stomach cancer. With this knowledge we may be able to stimulate pro-apoptotic p53 genes in many types of cancer that only contain mutated BAX genes and stimulate pro-apoptotic BAX genes in many types of cancer that only have mutated p53 genes. Of course if the BAK gene were stimulated, apoptosis could likely be induced in cells that have either or both p53 and BAX genes that have mutated within them.

Having the ability to identify unique genes on specific cancer cell chromosomes is an ability that gives humanity great control. Of course knowing the genes that cause apoptosis and the genes that commonly mutate to cause cancer does not mean much together without realizing how the knowledge of both together may help us to more easily stimulate apoptosis in cancer cells. Utilizing our understanding of apoptosis genetics and cancer gene mutations we may be able to modify and effectively activate the apoptosis genes in the cancer cell chromosomes by reprogramming a retrovirus to promote apoptosis as it identifies and activates certain pro-apoptotic genes, while deactivating other genes in the cancer cells that normally inhibit apoptosis! Each type of cancer has many genetic variations, so each type of cancer will likely have many different cures. I have compiled the majority of my hypothesis data in an effort to help the world to recognize that a curative process, but not just a single cure for cancer is possible and may exist. Of course simply identifying apoptosis genes within cancer cells does not do much good if those genes cannot be modified. We should also note the ways in which apoptosis may be stimulated in order to take advantage and experiment with the ways that look most simple to reach that goal of apoptosis, of course.

Which Apoptosis Genes
Are Available In Each Cancer?

While there are many genes that have mutated in each type of cancer cell noted we must remember that there are many apoptosis genes within human DNA that may still be available to stimulate and inhibit apoptosis in certain cancer cells. Just because some of the human DNA in cancer cells have become mutated does not mean that the rest of the genes have stopped functioning. There are many apoptosis genes that are still likely available within each cancer cells type as cancer was just a group of normal human cells before they mutated and began to reproduce. My curative cancer treatment process is all about stimulating the right apoptosis genes in cancer cells to promote apoptosis so that the cancers can be naturally flushed from the body, hopefully avoiding any major side effects.

Bladder Cancer

Apoptosis Inhibitors – API1, API2, API3, API4, API5, AVEN, BCL2, BCL2A1, BCLX, BIRC6, BIRC7, CIAPIN1, FAIM, FAIM2, FAIM3, MCL1, NAIP, PPP1R13L, PPP1R15A, SYVN1, TRIAP1

Apoptosis Regulators – BCLG, BCLW, BFAR, CCAR1, CFLAR, PAWR

Apoptosis Modulators – MOAP1, PUMA

Apoptosis Facilitators – BCL2L10, BCL2L11, BCL2L13, BCL2L14

Apoptosis Inducers – AIFM1, AIFM2, APITD1, BAD, BAK, BAX, BCL2L15, BIK, BIM, BMF, BOK, CIDEB, HRK, MRPS30, NAIF1, NOXA, PDCD1, PDCD2, PDCD4, PDCD5, PDCD6, PDCD6IP, PDCD7, PDCD10, PDCD11, PERP, SIVA1, TP53AIP1

Apoptosis Receptor Ligands – FASLG, Lymphotoxin Alpha, APO3L, APO2L, PDCD1LG1, PDCD1LG2

Apoptosis Receptors – DR1, DR2, DR3, DR4, DR5

Apoptosis Associates – AATF, AATK, AEN, CASP1, CASP2, CASP3, CASP4, CASP5, CASP6, CASP7, CASP8, CASP9, CASP10, CASP14, DBNDD2, FADD, PIDD, THAP1, THAP2, THAP3, TRADD

Bone Cancer [Osteosarcoma]

Apoptosis Inhibitors – API1, API2, API3, API4, API5, AVEN, BCL2, BCL2A1, BCLX, BIRC6, BIRC7, CIAPIN1, FAIM, FAIM2, FAIM3, MCL1, NAIP, PPP1R13L, PPP1R15A, SYVN1, TRIAP1

Apoptosis Regulators – BCLG, BCLW, BFAR, CCAR1, CFLAR, PAWR

Apoptosis Modulators – MOAP1, PUMA

Apoptosis Facilitators – BCL2L10, BCL2L11, BCL2L13, BCL2L14

Apoptosis Inducers – AIFM1, AIFM2, APITD1, BAD, BAK, BAX, BCL2L15, BIK, BIM, BMF, BOK, CIDEB, HRK, MRPS30, NAIF1, NOXA, PDCD1, PDCD2, PDCD4, PDCD5, PDCD6, PDCD6IP, PDCD7, PDCD10, PDCD11, PERP, SIVA1, TP53AIP1

Apoptosis Receptor Ligands – FASLG, Lymphotoxin Alpha, APO3L, APO2L, PDCD1LG1, PDCD1LG2

Apoptosis Receptors – DR1, DR2, DR3, DR4, DR5

Apoptosis Associates – AATF, AATK, AEN, CASP1, CASP2, CASP3, CASP4, CASP5, CASP6, CASP7, CASP8, CASP9, CASP10, CASP14, DBNDD2, FADD, PIDD, THAP1, THAP2, THAP3, TRADD

Brain/CNS Cancer

Apoptosis Inhibitors – API1, API2, API3, API5, AVEN, BCL2, BCL2A1, BCLX, BIRC6, BIRC7, CIAPIN1, FAIM, FAIM2, FAIM3, MCL1, NAIP, PPP1R13L, PPP1R15A, SYVN1, TRIAP1

Apoptosis Regulators – BCLG, BCLW, BFAR, CCAR1, CFLAR, PAWR

Apoptosis Modulators – MOAP1, PUMA

Apoptosis Facilitators – BCL2L10, BCL2L11, BCL2L13, BCL2L14

Apoptosis Inducers – AIFM1, AIFM2, APITD1, BAD, BAK, BAX, BCL2L15, BIK, BIM, BMF, BOK, CIDEB, HRK, MRPS30, NAIF1, NOXA, PDCD1, PDCD2, PDCD4, PDCD5, PDCD6, PDCD6IP, PDCD7, PDCD10, PDCD11, PERP, SIVA1, TP53AIP1

Apoptosis Receptor Ligands – FASLG, Lymphotoxin Alpha, APO3L, APO2L, PDCD1LG1, PDCD1LG2

Apoptosis Receptors – DR1, DR2, DR3, DR4, DR5

Apoptosis Associates – AATF, AATK, AEN, CASP1, CASP2, CASP3, CASP4, CASP5, CASP6, CASP7, CASP8, CASP9, CASP10, CASP14, DBNDD2, FADD, PIDD, THAP1, THAP2, THAP3, TRADD

Breast Cancer

Apoptosis Inhibitors – API1, API2, API3, API4, API5, AVEN, BCL2, BCL2A1, BCLX, BIRC6, BIRC7, CIAPIN1, FAIM, FAIM2, FAIM3, MCL1, NAIP, PPP1R13L, PPP1R15A, SYVN1, TRIAP1

Apoptosis Regulators – BCLG, BCLW, BFAR, CCAR1, CFLAR, PAWR

Apoptosis Modulators – MOAP1, PUMA

Apoptosis Facilitators – BCL2L10, BCL2L11, BCL2L13, BCL2L14

Apoptosis Inducers – AIFM1, AIFM2, APITD1, BAD, BAK, BAX, BCL2L15, BIK, BIM, BMF, BOK, CIDEB, HRK, MRPS30, NAIF1, NOXA, PDCD1, PDCD2, PDCD4, PDCD5, PDCD6, PDCD6IP, PDCD7, PDCD10, PDCD11, PERP, SIVA1, TP53AIP1

Apoptosis Receptor Ligands – FASLG, Lymphotoxin Alpha, APO3L, APO2L, PDCD1LG1, PDCD1LG2

Apoptosis Receptors – DR1, DR2, DR3, DR4, DR5

Apoptosis Associates – AATF, AATK, AEN, CASP1, CASP2, CASP3, CASP4, CASP5, CASP6, CASP7, CASP8, CASP9, CASP10, CASP14, DBNDD2, FADD, PIDD, THAP1, THAP2, THAP3, TRADD

Colorectal Cancer

Apoptosis Inhibitors – API1, API2, API3, API4, API5, AVEN, BCL2A1, BCLX, BIRC6, BIRC7, CIAPIN1, FAIM, FAIM2, FAIM3, MCL1, NAIP, PPP1R13L, PPP1R15A, SYVN1, TRIAP1

Apoptosis Regulators – BCLG, BCLW, BFAR, CCAR1, CFLAR, PAWR

Apoptosis Modulators – MOAP1, PUMA

Apoptosis Facilitators – BCL2L10, BCL2L11, BCL2L13, BCL2L14

Apoptosis Inducers – AIFM1, AIFM2, APITD1, BAD, BAK, BCL2L15, BIK, BIM, BMF, BOK, CIDEB, HRK,

MRPS30, NAIF1, NOXA, PDCD1, PDCD2, PDCD4, PDCD5, PDCD6, PDCD6IP, PDCD7, PDCD10, PDCD11, PERP, SIVA1, TP53AIP1

Apoptosis Receptor Ligands – FASLG, Lymphotoxin Alpha, APO3L, APO2L, PDCD1LG1, PDCD1LG2

Apoptosis Receptors – DR1, DR2, DR3, DR4

Apoptosis Associates – AATF, AATK, AEN, CASP1, CASP2, CASP3, CASP4, CASP5, CASP6, CASP7, CASP8, CASP9, CASP10, CASP14, DBNDD2, FADD, PIDD, THAP1, THAP2, THAP3, TRADD

Kidney Cancer

Apoptosis Inhibitors – API1, API2, API3, API4, API5, AVEN, BCL2, BCL2A1, BCLX, BIRC6, BIRC7, CIAPIN1, FAIM, FAIM2, FAIM3, MCL1, NAIP, PPP1R13L, PPP1R15A, SYVN1, TRIAP1

Apoptosis Regulators – BCLG, BCLW, BFAR, CCAR1, CFLAR, PAWR, TP53

Apoptosis Modulators – MOAP1, PUMA

Apoptosis Facilitators – BCL2L10, BCL2L11, BCL2L13, BCL2L14

Apoptosis Inducers – AIFM1, AIFM2, APITD1, BAD, BAK, BAX, BCL2L15, BIK, BIM, BMF, BOK, CIDEB, HRK, MRPS30, NAIF1, NOXA, PDCD1, PDCD2,

PDCD4, PDCD5, PDCD6, PDCD6IP, PDCD7, PDCD10, PDCD11, PERP, SIVA1, TP53AIP1

Apoptosis Receptor Ligands – FASLG, Lymphotoxin Alpha, APO3L, APO2L, PDCD1LG1, PDCD1LG2

Apoptosis Receptors – DR1, DR2, DR3, DR4, DR5

Apoptosis Associates – AATF, AATK, AEN, CASP1, CASP2, CASP3, CASP4, CASP5, CASP6, CASP7, CASP8, CASP9, CASP10, CASP14, DBNDD2, FADD, PIDD, THAP1, THAP2, THAP3, TRADD

Leukemia

Acute Lymphocytic Leukemia

Apoptosis Inhibitors – API1, API2, API3, API4, API5, AVEN, BCL2, BCL2A1, BCLX, BIRC6, BIRC7, CIAPIN1, FAIM, FAIM2, FAIM3, MCL1, NAIP, PPP1R13L, PPP1R15A, SYVN1, TRIAP1

Apoptosis Regulators – BCLG, BCLW, BFAR, CCAR1, CFLAR, PAWR, TP53

Apoptosis Modulators – MOAP1, PUMA

Apoptosis Facilitators – BCL2L10, BCL2L11, BCL2L13, BCL2L14

Apoptosis Inducers – AIFM1, AIFM2, APITD1, BAD, BAK, BCL2L15, BIK, BIM, BMF, BOK, CIDEB, HRK, MRPS30, NAIF1, NOXA, PDCD1, PDCD2, PDCD4,

PDCD5, PDCD6, PDCD6IP, PDCD7, PDCD10, PDCD11, PERP, SIVA1, TP53AIP1

Apoptosis Receptor Ligands – FASLG, Lymphotoxin Alpha, APO3L, APO2L, PDCD1LG1, PDCD1LG2

Apoptosis Receptors – DR2, DR3, DR4, DR5

Apoptosis Associates – AATF, AATK, AEN, CASP1, CASP2, CASP3, CASP4, CASP5, CASP6, CASP7, CASP8, CASP9, CASP10, CASP14, DBNDD2, FADD, PIDD, THAP1, THAP2, THAP3, TRADD

Acute Myeloid Leukemia

Apoptosis Inhibitors – API1, API2, API3, API4, API5, AVEN, BCL2A1, BCLX, BIRC6, BIRC7, CIAPIN1, FAIM, FAIM2, FAIM3, MCL1, NAIP, PPP1R13L, PPP1R15A, SYVN1, TRIAP1

Apoptosis Regulators – BCLG, BCLW, BFAR, CCAR1, CFLAR, PAWR, TP53

Apoptosis Modulators – MOAP1, PUMA

Apoptosis Facilitators – BCL2L10, BCL2L11, BCL2L13, BCL2L14

Apoptosis Inducers – AIFM1, AIFM2, APITD1, BAD, BAK, BCL2L15, BIK, BIM, BMF, BOK, CIDEB, HRK, MRPS30, NAIF1, NOXA, PDCD1, PDCD2, PDCD4, PDCD5, PDCD6, PDCD6IP, PDCD7, PDCD10, PDCD11, PERP, SIVA1, TP53AIP1

Apoptosis Receptor Ligands – FASLG, Lymphotoxin Alpha, APO3L, APO2L, PDCD1LG1, PDCD1LG2

Apoptosis Receptors – DR1, DR2, DR3, DR4, DR5

Apoptosis Associates – AATF, AATK, AEN, CASP1, CASP2, CASP3, CASP4, CASP5, CASP6, CASP7, CASP8, CASP9, CASP10, CASP14, DBNDD2, FADD, PIDD, THAP1, THAP2, THAP3, TRADD

Chronic Lymphocytic Leukemia

Apoptosis Inhibitors – API1, API2, API3, API4, API5, AVEN, BCL2, BCL2A1, BCLX, BIRC6, BIRC7, CIAPIN1, FAIM, FAIM2, FAIM3, MCL1, NAIP, PPP1R13L, PPP1R15A, SYVN1, TRIAP1

Apoptosis Regulators – BCLG, BCLW, BFAR, CCAR1, CFLAR, PAWR

Apoptosis Modulators – MOAP1, PUMA

Apoptosis Facilitators – BCL2L10, BCL2L11, BCL2L13, BCL2L14

Apoptosis Inducers – AIFM1, AIFM2, APITD1, BAD, BAK, BCL2L15, BIK, BIM, BMF, BOK, CIDEB, HRK, MRPS30, NAIF1, NOXA, PDCD1, PDCD2, PDCD4, PDCD5, PDCD6, PDCD6IP, PDCD7, PDCD10, PDCD11, PERP, SIVA1, TP53AIP1

Apoptosis Receptor Ligands – FASLG, Lymphotoxin Alpha, APO3L, APO2L, PDCD1LG1, PDCD1LG2

Apoptosis Receptors – DR1, DR2, DR3, DR4, DR5

Apoptosis Associates – AATF, AATK, AEN, CASP1, CASP2, CASP3, CASP4, CASP5, CASP6, CASP7, CASP8, CASP9, CASP10, CASP14, DBNDD2, FADD, PIDD, THAP1, THAP2, THAP3, TRADD

Chronic Myeloid Leukemia

Apoptosis Inhibitors – API1, API2, API3, API4, API5, AVEN, BCL2, BCL2A1, BCLX, BIRC6, BIRC7, CIAPIN1, FAIM, FAIM2, FAIM3, MCL1, NAIP, PPP1R13L, PPP1R15A, SYVN1, TRIAP1

Apoptosis Regulators – BCLG, BCLW, BFAR, CCAR1, CFLAR, PAWR, TP53

Apoptosis Modulators – MOAP1, PUMA

Apoptosis Facilitators – BCL2L10, BCL2L11, BCL2L13, BCL2L14

Apoptosis Inducers – AIFM1, AIFM2, APITD1, BAD, BAK, BAX, BCL2L15, BIK, BIM, BMF, BOK, CIDEB, HRK, MRPS30, NAIF1, NOXA, PDCD1, PDCD2, PDCD4, PDCD5, PDCD6, PDCD6IP, PDCD7, PDCD10, PDCD11, PERP, SIVA1, TP53AIP1

Apoptosis Receptor Ligands – FASLG, Lymphotoxin Alpha, APO3L, APO2L, PDCD1LG1, PDCD1LG2

Apoptosis Receptors – DR1, DR2, DR3, DR4, DR5

Apoptosis Associates – AATF, AATK, AEN, CASP1, CASP2, CASP3, CASP4, CASP5, CASP6, CASP7, CASP8, CASP9, CASP10, CASP14, DBNDD2, FADD, PIDD, THAP1, THAP2, THAP3, TRADD

Eosinophilic Leukemia

Apoptosis Inhibitors – API1, API2, API3, API4, API5, AVEN, BCL2, BCL2A1, BCLX, BIRC6, BIRC7, CIAPIN1, FAIM, FAIM2, FAIM3, MCL1, NAIP, PPP1R13L, PPP1R15A, SYVN1, TRIAP1

Apoptosis Regulators – BCLG, BCLW, BFAR, CCAR1, CFLAR, PAWR

Apoptosis Modulators – MOAP1, PUMA

Apoptosis Facilitators – BCL2L10, BCL2L11, BCL2L13, BCL2L14

Apoptosis Inducers – AIFM1, AIFM2, APITD1, BAD, BAK, BAX, BCL2L15, BIK, BIM, BMF, BOK, CIDEB, HRK, MRPS30, NAIF1, NOXA, PDCD1, PDCD2, PDCD4, PDCD5, PDCD6, PDCD6IP, PDCD7, PDCD10, PDCD11, PERP, SIVA1, TP53AIP1

Apoptosis Receptor Ligands – FASLG, Lymphotoxin Alpha, APO3L, APO2L, PDCD1LG1, PDCD1LG2

Apoptosis Receptors – DR1, DR2, DR3, DR4, DR5

Apoptosis Associates – AATF, AATK, AEN, CASP1, CASP2, CASP3, CASP4, CASP5, CASP6, CASP7,

CASP8, CASP9, CASP10, CASP14, DBNDD2, FADD, PIDD, THAP1, THAP2, THAP3, TRADD

Liver Cancer

Apoptosis Inhibitors – API1, API2, API3, API5, AVEN, BCL2, BCL2A1, BCLX, BIRC6, BIRC7, CIAPIN1, FAIM, FAIM2, FAIM3, MCL1, NAIP, PPP1R13L, PPP1R15A, SYVN1, TRIAP1

Apoptosis Regulators – BCLG, BCLW, BFAR, CCAR1, CFLAR, PAWR

Apoptosis Modulators – MOAP1, PUMA

Apoptosis Facilitators – BCL2L10, BCL2L11, BCL2L13, BCL2L14

Apoptosis Inducers – AIFM1, AIFM2, APITD1, BAD, BAK, BAX, BCL2L15, BIK, BIM, BMF, BOK, CIDEB, HRK, MRPS30, NAIF1, NOXA, PDCD1, PDCD2, PDCD4, PDCD5, PDCD6, PDCD6IP, PDCD7, PDCD10, PDCD11, PERP, SIVA1, TP53AIP1

Apoptosis Receptor Ligands – FASLG, Lymphotoxin Alpha, APO3L, APO2L, PDCD1LG1, PDCD1LG2

Apoptosis Receptors – DR1, DR2, DR3, DR4, DR5

Apoptosis Associates – AATF, AATK, AEN, CASP1, CASP2, CASP3, CASP4, CASP5, CASP6, CASP7, CASP8, CASP9, CASP10, CASP14, DBNDD2, FADD, PIDD, THAP1, THAP2, THAP3, TRADD

Lung Cancer

Apoptosis Inhibitors – API1, API2, API3, API5, AVEN, BCL2, BCL2A1, BCLX, BIRC6, BIRC7, CIAPIN1, FAIM, FAIM2, FAIM3, MCL1, NAIP, PPP1R13L, PPP1R15A, SYVN1, TRIAP1

Apoptosis Regulators – BCLG, BCLW, BFAR, CCAR1, CFLAR, PAWR

Apoptosis Modulators – MOAP1, PUMA

Apoptosis Facilitators – BCL2L10, BCL2L11, BCL2L13, BCL2L14

Apoptosis Inducers – AIFM1, AIFM2, APITD1, BAD, BAK, BAX, BCL2L15, BIK, BIM, BMF, BOK, CIDEB, HRK, MRPS30, NAIF1, NOXA, PDCD1, PDCD2, PDCD4, PDCD5, PDCD6, PDCD6IP, PDCD7, PDCD10, PDCD11, PERP, SIVA1, TP53AIP1

Apoptosis Receptor Ligands – FASLG, Lymphotoxin Alpha, APO3L, APO2L, PDCD1LG1, PDCD1LG2

Apoptosis Receptors – DR1, DR2, DR3, DR4, DR5

Apoptosis Associates – AATF, AATK, AEN, CASP1, CASP2, CASP3, CASP4, CASP5, CASP6, CASP7, CASP8, CASP9, CASP10, CASP14, DBNDD2, FADD, PIDD, THAP1, THAP2, THAP3, TRADD

Hodgkin Lymphoma

Apoptosis Inhibitors – API1, API2, API3, API5, AVEN, BCL2A1, BCLX, BIRC6, BIRC7, CIAPIN1, FAIM, FAIM2, FAIM3, MCL1, NAIP, PPP1R13L, PPP1R15A, SYVN1, TRIAP1

Apoptosis Regulators – BCLG, BCLW, BFAR, CCAR1, CFLAR, PAWR, TP53

Apoptosis Modulators – MOAP1, PUMA

Apoptosis Facilitators – BCL2L10, BCL2L11, BCL2L13, BCL2L14

Apoptosis Inducers – AIFM1, AIFM2, APITD1, BAD, BAK, BAX, BCL2L15, BIK, BIM, BMF, BOK, CIDEB, HRK, MRPS30, NAIF1, NOXA, PDCD1, PDCD2, PDCD4, PDCD5, PDCD6, PDCD6IP, PDCD7, PDCD10, PDCD11, PERP, SIVA1, TP53AIP1

Apoptosis Receptor Ligands – FASLG, Lymphotoxin Alpha, APO3L, APO2L, PDCD1LG1, PDCD1LG2

Apoptosis Receptors – DR1, DR2, DR3, DR4, DR5

Apoptosis Associates – AATF, AATK, AEN, CASP1, CASP2, CASP3, CASP4, CASP5, CASP6, CASP7, CASP8, CASP9, CASP10, CASP14, DBNDD2, FADD, PIDD, THAP1, THAP2, THAP3, TRADD

Non-Hodgkin Lymphoma

Apoptosis Inhibitors – API1, API2, API3, API4, API5, AVEN, BCL2A1, BCLX, BIRC6, BIRC7, CIAPIN1, FAIM, FAIM2, FAIM3, MCL1, NAIP, PPP1R13L, PPP1R15A, SYVN1, TRIAP1

Apoptosis Regulators – BCLG, BCLW, BFAR, CCAR1, CFLAR, PAWR

Apoptosis Modulators – MOAP1, PUMA

Apoptosis Facilitators – BCL2L10, BCL2L11, BCL2L13, BCL2L14

Apoptosis Inducers – AIFM1, AIFM2, APITD1, BAD, BAK, BAX, BCL2L15, BIK, BIM, BMF, BOK, CIDEB, HRK, MRPS30, NAIF1, NOXA, PDCD1, PDCD2, PDCD4, PDCD5, PDCD6, PDCD6IP, PDCD7, PDCD10, PDCD11, PERP, SIVA1, TP53AIP1

Apoptosis Receptor Ligands – FASLG, Lymphotoxin Alpha, APO3L, APO2L, PDCD1LG1, PDCD1LG2

Apoptosis Receptors – DR1, DR2, DR3, DR4, DR5

Apoptosis Associates – AATF, AATK, AEN, CASP1, CASP2, CASP3, CASP4, CASP5, CASP6, CASP7, CASP8, CASP9, CASP10, CASP14, DBNDD2, FADD, PIDD, THAP1, THAP2, THAP3, TRADD

Myeloma

Apoptosis Inhibitors – API1, API2, API3, API4, API5, AVEN, BCL2, BCL2A1, BCLX, BIRC6, BIRC7, CIAPIN1, FAIM, FAIM2, FAIM3, MCL1, NAIP, PPP1R13L, PPP1R15A, SYVN1, TRIAP1

Apoptosis Regulators – BCLG, BCLW, BFAR, CCAR1, CFLAR, PAWR, TP53

Apoptosis Modulators – MOAP1, PUMA

Apoptosis Facilitators – BCL2L10, BCL2L11, BCL2L13, BCL2L14

Apoptosis Inducers – AIFM1, AIFM2, APITD1, BAD, BAK, BAX, BCL2L15, BIK, BIM, BMF, BOK, CIDEB, HRK, MRPS30, NAIF1, NOXA, PDCD1, PDCD2, PDCD4, PDCD5, PDCD6, PDCD6IP, PDCD7, PDCD10, PDCD11, PERP, SIVA1, TP53AIP1

Apoptosis Receptor Ligands – FASLG, Lymphotoxin Alpha, APO3L, APO2L, PDCD1LG1, PDCD1LG2

Apoptosis Receptors – DR1, DR2, DR3, DR4, DR5

Apoptosis Associates – AATF, AATK, AEN, CASP1, CASP2, CASP3, CASP4, CASP5, CASP6, CASP7, CASP8, CASP9, CASP10, CASP14, DBNDD2, FADD, PIDD, THAP1, THAP2, THAP3, TRADD

Ovarian Cancer

Apoptosis Inhibitors – API1, API2, API3, API4, API5, AVEN, BCL2A1, BCLX, BIRC6, BIRC7, CIAPIN1, FAIM, FAIM2, FAIM3, MCL1, NAIP, PPP1R13L, PPP1R15A, SYVN1, TRIAP1

Apoptosis Regulators – BCLG, BCLW, BFAR, CCAR1, CFLAR, PAWR

Apoptosis Modulators – MOAP1, PUMA

Apoptosis Facilitators – BCL2L10, BCL2L11, BCL2L13, BCL2L14

Apoptosis Inducers – AIFM1, AIFM2, APITD1, BAD, BAK, BAX, BCL2L15, BIK, BIM, BMF, BOK, CIDEB, HRK, MRPS30, NAIF1, NOXA, PDCD1, PDCD2, PDCD4, PDCD5, PDCD6, PDCD6IP, PDCD7, PDCD10, PDCD11, PERP, SIVA1, TP53AIP1

Apoptosis Receptor Ligands – FASLG, Lymphotoxin Alpha, APO3L, APO2L, PDCD1LG1, PDCD1LG2

Apoptosis Receptors – DR1, DR2, DR3, DR4, DR5

Apoptosis Associates – AATF, AATK, AEN, CASP1, CASP2, CASP3, CASP4, CASP5, CASP6, CASP7, CASP8, CASP9, CASP10, CASP14, DBNDD2, FADD, PIDD, THAP1, THAP2, THAP3, TRADD

Pancreatic Cancer

Apoptosis Inhibitors – API1, API2, API3, API4, API5, AVEN, BCL2, BCL2A1, BCLX, BIRC6, BIRC7, CIAPIN1, FAIM, FAIM2, FAIM3, MCL1, NAIP, PPP1R13L, PPP1R15A, SYVN1, TRIAP1

Apoptosis Regulators – BCLG, BCLW, BFAR, CCAR1, CFLAR, PAWR, TP53

Apoptosis Modulators – MOAP1, PUMA

Apoptosis Facilitators – BCL2L10, BCL2L11, BCL2L13, BCL2L14

Apoptosis Inducers – AIFM1, AIFM2, APITD1, BAD, BAK, BCL2L15, BIK, BIM, BMF, BOK, CIDEB, HRK, MRPS30, NAIF1, NOXA, PDCD1, PDCD2, PDCD4, PDCD5, PDCD6, PDCD6IP, PDCD7, PDCD10, PDCD11, PERP, SIVA1, TP53AIP1

Apoptosis Receptor Ligands – FASLG, Lymphotoxin Alpha, APO3L, APO2L, PDCD1LG1, PDCD1LG2

Apoptosis Receptors – DR1, DR2, DR3, DR4, DR5

Apoptosis Associates – AATF, AATK, AEN, CASP1, CASP2, CASP3, CASP4, CASP5, CASP6, CASP7, CASP8, CASP9, CASP10, CASP14, DBNDD2, FADD, PIDD, THAP1, THAP2, THAP3, TRADD

Prostate Cancer

Apoptosis Inhibitors – API1, API2, API3, API4, API5, AVEN, BCL2, BCL2A1, BCLX, BIRC6, BIRC7, CIAPIN1, FAIM, FAIM2, FAIM3, MCL1, NAIP, PPP1R13L, PPP1R15A, SYVN1, TRIAP1

Apoptosis Regulators – BCLG, BCLW, BFAR, CCAR1, CFLAR, PAWR

Apoptosis Modulators – MOAP1, PUMA

Apoptosis Facilitators – BCL2L10, BCL2L11, BCL2L13, BCL2L14

Apoptosis Inducers – AIFM1, AIFM2, APITD1, BAD, BAK, BAX, BCL2L15, BIK, BIM, BMF, BOK, CIDEB, HRK, MRPS30, NAIF1, NOXA, PDCD1, PDCD2, PDCD4, PDCD5, PDCD6, PDCD6IP, PDCD7, PDCD10, PDCD11, PERP, SIVA1, TP53AIP1

Apoptosis Receptor Ligands – FASLG, Lymphotoxin Alpha, APO3L, APO2L, PDCD1LG1, PDCD1LG2

Apoptosis Receptors – DR1, DR2, DR3, DR4, DR5

Apoptosis Associates – AATF, AATK, AEN, CASP1, CASP2, CASP3, CASP4, CASP5, CASP6, CASP7, CASP8, CASP9, CASP10, CASP14, DBNDD2, FADD, PIDD, THAP1, THAP2, THAP3, TRADD

Melanoma Skin Cancer

Apoptosis Inhibitors – API1, API2, API3, API4, API5, AVEN, BCL2, BCL2A1, BCLX, BIRC6, BIRC7, CIAPIN1, FAIM, FAIM2, FAIM3, MCL1, NAIP, PPP1R13L, PPP1R15A, SYVN1, TRIAP1

Apoptosis Regulators – BCLG, BCLW, BFAR, CCAR1, CFLAR, PAWR

Apoptosis Modulators – MOAP1, PUMA

Apoptosis Facilitators – BCL2L10, BCL2L11, BCL2L13, BCL2L14

Apoptosis Inducers – AIFM1, AIFM2, APITD1, BAD, BAK, BAX, BCL2L15, BIK, BIM, BMF, BOK, CIDEB, HRK, MRPS30, NAIF1, NOXA, PDCD1, PDCD2, PDCD4, PDCD5, PDCD6, PDCD6IP, PDCD7, PDCD10, PDCD11, PERP, SIVA1, TP53AIP1

Apoptosis Receptor Ligands – FASLG, Lymphotoxin Alpha, APO3L, APO2L, PDCD1LG1, PDCD1LG2

Apoptosis Receptors – DR1, DR2, DR3, DR4, DR5

Apoptosis Associates – AATF, AATK, AEN, CASP1, CASP2, CASP3, CASP4, CASP5, CASP6, CASP7, CASP8, CASP9, CASP10, CASP14, DBNDD2, FADD, PIDD, THAP1, THAP2, THAP3, TRADD

Basal Cell Carcinoma Skin Cancer

Apoptosis Inhibitors – API1, API2, API3, API4, API5, AVEN, BCL2, BCL2A1, BCLX, BIRC6, BIRC7, CIAPIN1, FAIM, FAIM2, FAIM3, MCL1, NAIP, PPP1R13L, PPP1R15A, SYVN1, TRIAP1

Apoptosis Regulators – BCLG, BCLW, BFAR, CCAR1, CFLAR, PAWR, TP53

Apoptosis Modulators – MOAP1, PUMA

Apoptosis Facilitators – BCL2L10, BCL2L11, BCL2L13, BCL2L14

Apoptosis Inducers – AIFM1, AIFM2, APITD1, BAD, BAK, BAX, BCL2L15, BIK, BIM, BMF, BOK, CIDEB, HRK, MRPS30, NAIF1, NOXA, PDCD1, PDCD2, PDCD4, PDCD5, PDCD6, PDCD6IP, PDCD7, PDCD10, PDCD11, PERP, SIVA1, TP53AIP1

Apoptosis Receptor Ligands – FASLG, Lymphotoxin Alpha, APO3L, APO2L, PDCD1LG1, PDCD1LG2

Apoptosis Receptors – DR1, DR2, DR3, DR4, DR5

Apoptosis Associates – AATF, AATK, AEN, CASP1, CASP2, CASP3, CASP4, CASP5, CASP6, CASP7, CASP8, CASP9, CASP10, CASP14, DBNDD2, FADD, PIDD, THAP1, THAP2, THAP3, TRADD

Stomach Cancer

Apoptosis Inhibitors – API1, API2, API3, API4, API5, AVEN, BCL2, BCL2A1, BCLX, BIRC6, BIRC7, CIAPIN1, FAIM, FAIM2, FAIM3, MCL1, NAIP, PPP1R13L, PPP1R15A, SYVN1, TRIAP1

Apoptosis Regulators – BCLG, BCLW, BFAR, CCAR1, CFLAR, PAWR

Apoptosis Modulators – MOAP1, PUMA

Apoptosis Facilitators – BCL2L10, BCL2L11, BCL2L13, BCL2L14

Apoptosis Inducers – AIFM1, AIFM2, APITD1, BAD, BAK, BCL2L15, BIK, BIM, BMF, BOK, CIDEB, HRK, MRPS30, NAIF1, NOXA, PDCD1, PDCD2, PDCD4, PDCD5, PDCD6, PDCD6IP, PDCD7, PDCD10, PDCD11, PERP, SIVA1, TP53AIP1

Apoptosis Receptor Ligands – FASLG, Lymphotoxin Alpha, APO3L, APO2L, PDCD1LG1, PDCD1LG2

Apoptosis Receptors – DR1, DR2, DR3, DR4, DR5

Apoptosis Associates – AATF, AATK, AEN, CASP1, CASP2, CASP3, CASP4, CASP5, CASP6, CASP7, CASP8, CASP9, CASP10, CASP14, DBNDD2, FADD, PIDD, THAP1, THAP2, THAP3, TRADD

Testicular Cancer

Apoptosis Inhibitors – API1, API2, API3, API4, API5, AVEN, BCL2, BCL2A1, BCLX, BIRC6, BIRC7, CIAPIN1, FAIM, FAIM2, FAIM3, MCL1, NAIP, PPP1R13L, PPP1R15A, SYVN1, TRIAP1

Apoptosis Regulators – BCLG, BCLW, BFAR, CCAR1, CFLAR, PAWR, TP53

Apoptosis Modulators – MOAP1, PUMA

Apoptosis Facilitators – BCL2L10, BCL2L11, BCL2L13, BCL2L14

Apoptosis Inducers – AIFM1, AIFM2, APITD1, BAD, BAK, BAX, BCL2L15, BIK, BIM, BMF, BOK, CIDEB, HRK, MRPS30, NAIF1, NOXA, PDCD1, PDCD2, PDCD4, PDCD5, PDCD6, PDCD6IP, PDCD7, PDCD10, PDCD11, PERP, SIVA1, TP53AIP1

Apoptosis Receptor Ligands – FASLG, Lymphotoxin Alpha, APO3L, APO2L, PDCD1LG1, PDCD1LG2

Apoptosis Receptors – DR1, DR2, DR3, DR4, DR5

Apoptosis Associates – AATF, AATK, AEN, CASP1, CASP2, CASP3, CASP4, CASP5, CASP6, CASP7, CASP8, CASP9, CASP10, CASP14, DBNDD2, FADD, PIDD, THAP1, THAP2, THAP3, TRADD

Thyroid Cancer

Apoptosis Inhibitors – API1, API2, API3, API4, API5, AVEN, BCL2, BCL2A1, BCLX, BIRC6, BIRC7, CIAPIN1, FAIM, FAIM2, FAIM3, MCL1, NAIP, PPP1R13L, PPP1R15A, SYVN1, TRIAP1

Apoptosis Regulators – BCLG, BCLW, BFAR, CCAR1, CFLAR, PAWR, TP53

Apoptosis Modulators – MOAP1, PUMA

Apoptosis Facilitators – BCL2L10, BCL2L11, BCL2L13, BCL2L14

Apoptosis Inducers – AIFM1, AIFM2, APITD1, BAD, BAK, BAX, BCL2L15, BIK, BIM, BMF, BOK, CIDEB, HRK, MRPS30, NAIF1, NOXA, PDCD1, PDCD2, PDCD4, PDCD5, PDCD6, PDCD6IP, PDCD7, PDCD10, PDCD11, PERP, SIVA1, TP53AIP1

Apoptosis Receptor Ligands – FASLG, Lymphotoxin Alpha, APO3L, APO2L, PDCD1LG1, PDCD1LG2

Apoptosis Receptors – DR1, DR2, DR3, DR4, DR5

Apoptosis Associates – AATF, AATK, AEN, CASP1, CASP2, CASP3, CASP4, CASP5, CASP6, CASP7, CASP8, CASP9, CASP10, CASP14, DBNDD2, FADD, PIDD, THAP1, THAP2, THAP3, TRADD

Uterine Cancer

Apoptosis Inhibitors – API1, API2, API3, API4, API5, AVEN, BCL2, BCL2A1, BCLX, BIRC6, BIRC7, CIAPIN1, FAIM, FAIM2, FAIM3, MCL1, NAIP, PPP1R13L, PPP1R15A, SYVN1, TRIAP1

Apoptosis Regulators – BCLG, BCLW, BFAR, CCAR1, CFLAR, PAWR

Apoptosis Modulators – MOAP1, PUMA

Apoptosis Facilitators – BCL2L10, BCL2L11, BCL2L13, BCL2L14

Apoptosis Inducers – AIFM1, AIFM2, APITD1, BAD, BAK, BAX, BCL2L15, BIK, BIM, BMF, BOK, CIDEB, HRK, MRPS30, NAIF1, NOXA, PDCD1, PDCD2, PDCD4, PDCD5, PDCD6, PDCD6IP, PDCD7, PDCD10, PDCD11, PERP, SIVA1, TP53AIP1

Apoptosis Receptor Ligands – FASLG, Lymphotoxin Alpha, APO3L, APO2L, PDCD1LG1, PDCD1LG2

Apoptosis Receptors – DR1, DR2, DR3, DR4, DR5

Apoptosis Associates – AATF, AATK, AEN, CASP1, CASP2, CASP3, CASP4, CASP5, CASP6, CASP7, CASP8, CASP9, CASP10, CASP14, DBNDD2, FADD, PIDD, THAP1, THAP2, THAP3, TRADD

How Can Apoptosis Be Stimulated In Cancer Cells?

"BAK overexpression mediates p53-independent apoptosis inducing effects on human gastric cancer cells", is a 2004 study that showed that by using a eukaryotic vector apoptosis could be induced in stomach cancer cells through the overexpression of the BAK gene. The BAK gene is a pro-apoptotic member of the Bcl-2 family and causes apoptosis to occur independently of the p53 pathway. As BAK was stimulated, anti-apoptotic genes Bcl-2 and Bcl-XL were inhibited. The results of this study showed that BAK overexpression promotes apoptosis and inhibits anti-apoptotic genes in gastric cancer cells independently of p53 [212].

"Opposite role of BAX and BCL-2 in the anti-tumoral responses of the immune system", a 2004 study, demonstrated the opposing roles of the genes regarding apoptosis in tumor cells. In vitro transfected tumor cells with increased levels of BAX showed that apoptosis was promoted in those cells. Whereas in vitro transfected tumor cells with increased levels of Bcl-2 showed that apoptosis was inhibited. This study also noted that in vivo transfected cell lines showed no growth with BAX, but also showed that large tumors formed with Bcl-2 [213].

A study performed in 2007, "E2F-1 induces melanoma cell apoptosis via PUMA up-regulation and BAX translocation", showed that the E2F-1 gene induces apoptosis in melanoma cells as PUMA genes are activated, and BAX is activated. PUMA and subsequent BAX activation contribute to E2F-1 induced apoptosis, although the PUMA gene is sometimes lost or mutated in cancer cells. If the PUMA gene is mutated or deleted in the cancer cells then E2F-1 cannot indirectly stimulate BAX properly to initiate apoptosis [214]. Of course it should be noted that E2F-1 can induce apoptosis through both the p53 apoptotic pathway, as well as through p53 independent apoptotic pathways.

"Proapoptotic BAX and BAK: A Requisite Gateway to Mitochondrial Dysfunction and Death", showed that the BAX or BAK gene can initiate mitochondrial dysfunction and cell death. During the study the hypothesis was confirmed that through retroviral stimulation of DR1, the tBID gene must be stimulated for BAX or BAK activated mitochondrial dysfunction and cell death in murine embryonic fibroblasts and hepatocytes, as tBID is caspase activated BID [215].

The 2003, "BAX and BAK can localize to the endoplasmic reticulum to initiate apoptosis"; study showed that the BAX and BAK genes exist in both the endoplasmic reticulum of a cell and the mitochondria membrane. As a result of these genes' dual locations, BAX or BAK apoptosis may be stimulated in either the endoplasmic reticulum or the mitochondria of a eukaryotic cell [216].

Intrinsic Apoptosis

[1p]

Intrinsic apoptosis is still possible when the p53 gene has mutated. It may be the most beneficial type of apoptosis when trying to destroy the cancer cells and flush them from the body's systems without harming any other areas of the body. I believe that both intrinsic and extrinsic methods should be tested to see which truly does work most effectively. I also expect that extrinsic apoptosis, being more direct and having fewer variables, will likely be more effective to induce into cancer cells. Fewer steps are needed to modify a vector to reach the goal of extrinsic apoptosis in cancer cells, which should be more ideal than the extra vector modification steps that will be needed to stimulate intrinsic apoptosis in various different cancer types.

Extrinsic apoptosis appears to be much more easily applied to all of the cancer types that I have noted in this book. As the proper apoptosis receptor ligands are stimulated and as a result the coordinating apoptosis receptor activated, the proper caspases will be released to destroy the cell which will then be naturally flushed and removed from the body. That sounds simple enough to me, but how can we activate these apoptosis receptor ligands to begin the process of extrinsic apoptosis? The key will be to find the most effective genetic vector/s for the project that can be modified to stimulate the right genes in a proper fashion.

Extrinsic Apoptosis

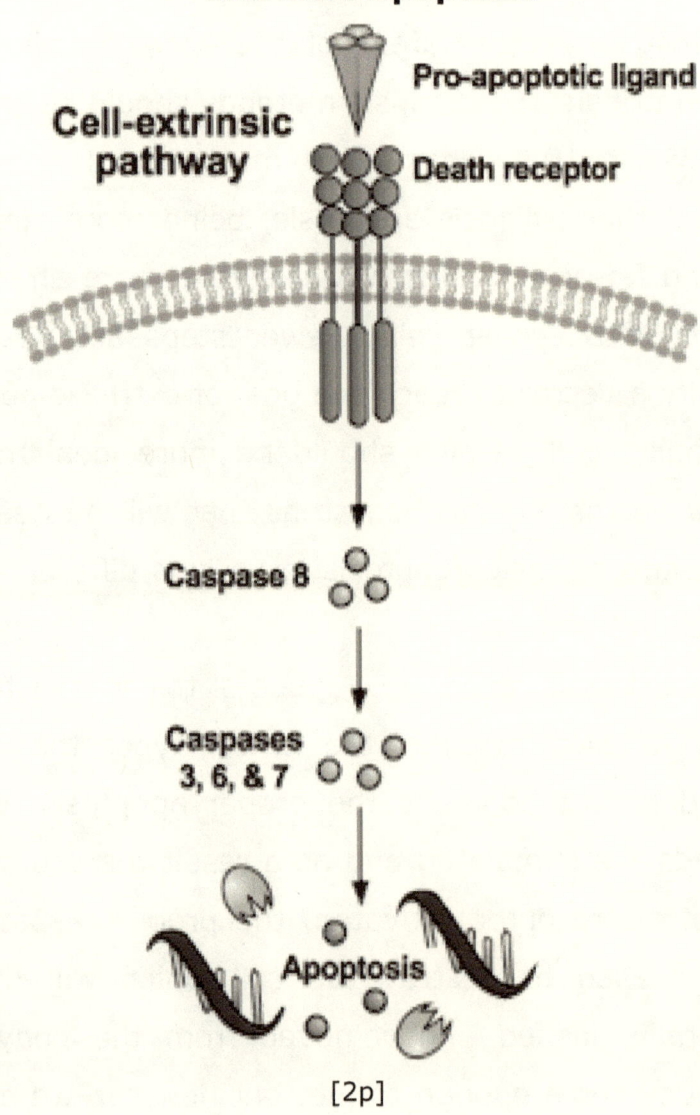

[2p]

A genetically modified virus can be modified to eliminate specific unwanted cells. Retroviruses are RNA viruses that naturally function to change the DNA of the cell types that they attack by inserting mRNA into a cell's DNA through reverse transcription, effectively changing the genetic DNA structure of the host cell that it is attacking. Of course, a genetically modified retrovirus could be utilized to alter specific genes as the retrovirus mRNA is inserted on one side on a DNA strand to effectively activate or inhibit specific genes to cause apoptosis in the cells which are targeted. A retrovirus can be reprogrammed to seek out specific cell types in the body. Once the specific cell type is found by a retrovirus it is absorbed by the cell receptors then through the cell plasma membrane, via endocytosis the virus is inserted into the cell cytoplasm where it begins uncoating. The virus' mRNA is then absorbed by the cell nucleus. As the retrovirus progresses it separates the cell DNA through reverse transcription creating two DNA chains out of one.

General Retrovirus Replication Cycle

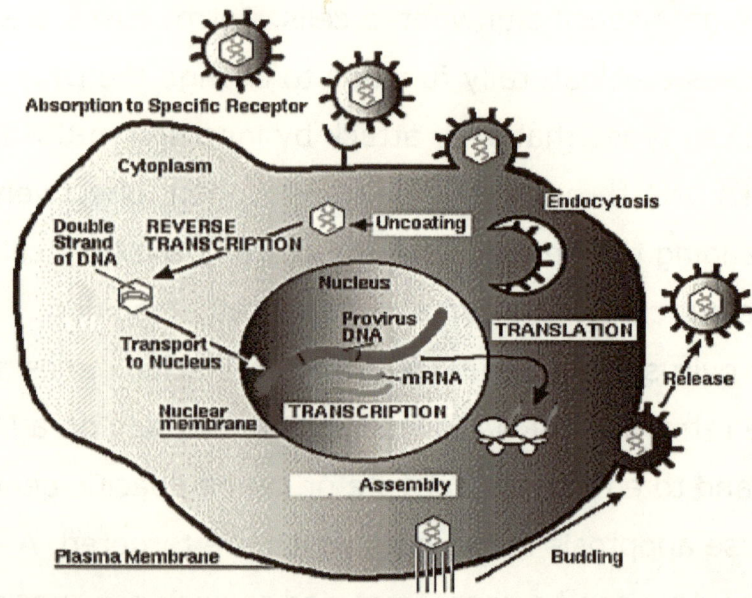

Retrovirus replication

[3p]

Retrovirus Pictorial Diagram

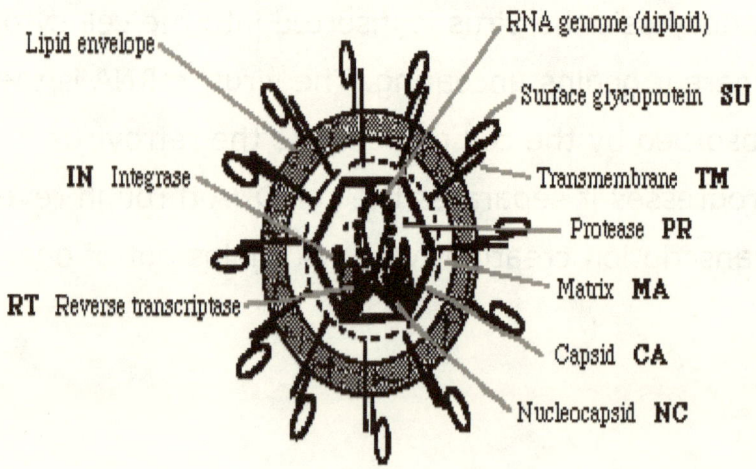

[4p]

Retrovirus Genomic Diagram

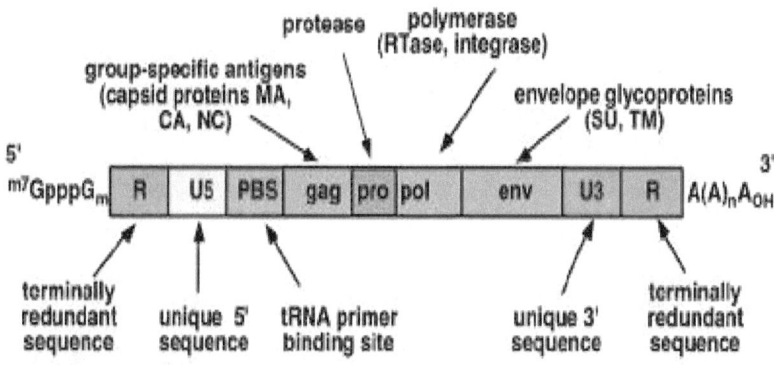

The Retrovirus Genome
(plus strand RNA)

[5p]

All retroviruses contain four common genes, which include the GAG, PRO, POL, and ENV genes that exist between the long terminal repeats at each end of the genetic sequence that have a part in the integration of the viral RNA into the DNA of the host cell. The GAG gene encodes internal structural proteins of the virus as this gene is proteolytically processed into a mature matrix [MA] protein that lines the virus envelope, a capsid [CA] protein which is the most abundant viral protein that protects the core, a nucleocapsid [NC] protein that forms the core and the genome, and sometimes other proteins of various functions that are designated by number. The PRO gene encodes the viral protease [PR] protein that acts late in the assembly of the viral particle and is vital for the cleavage of the GAG gene during maturation. The POL gene encodes the reverse transcriptase enzyme which reverse transcribes the RNA genome as it contains both, DNA polymerase and associated RNase H activities, as well as the integrase [IN] protein that is necessary for the integration of the provirus. The ENV gene encodes the surface glycoprotein [SU] which is the major virus antigen of the outer envelope glycoprotein, as well as encoding the transmembrane [TM] protein of the virion which is the inner component of the mature envelope glycoprotein. The ENV gene proteins form a complex that interacts with specific cellular receptor proteins as the interaction ultimately leads to fusion of the viral membrane with the cell membrane [217, 218].

I propose that a retrovirus be modified to target specific cancer cells in order to promote apoptosis within them by stimulating or inhibiting certain apoptosis genes noted throughout my research. I believe that a type C retrovirus may be the most beneficial type on which to begin genetically modifying. The GAG and POL retrovirus genes would need to be modified to promote apoptosis within the cells via p53 or independently of the p53 gene by stimulating and/or inhibiting various apoptotic genes. The GAG gene MA and NC proteins would be modified to activate or inhibit specific apoptotic genes, while the POL gene IN protein would be modified to integrate into the cellular DNA at a specific point where apoptosis genes may be stimulated. If the retrovirus GAG and POL genes are initially modified to promote p53 and p53 independent apoptosis to begin with in separate samples, those viruses could then be reproduced in laboratories. Once those types of modified retroviruses are available the ENV gene proteins, SU and TM, could then be genetically modified to target specific cancer cells depending of the genetic cancer type and not necessarily only T-cells or the cell type the particular retrovirus normally targets.

As a retrovirus is modified to promote p53, or p53 independent intrinsic apoptosis it may be utilized to target multiple cancer types as so many organs, tissue, and cells in the human body share many of the same types of apoptotic genes. As a result of having such genetically modified retroviruses available, apoptosis may be accomplished within various cancer cells without trying to stimulate already mutated apoptotic genes such as p53 or BAX. Of course, there are many other apoptotic genes that may be stimulated or inhibited within cancer cells. It may be easier to simply use a retrovirus for extrinsic apoptosis by stimulating death receptor ligands in the cancer cell to promote p53 independent apoptosis through activation of specific death receptors most of which are not noted as mutated. After which the modified retrovirus could be reproduced then its ENV proteins could be modified to target specific cancer cell types and not just the cell types that the retrovirus normally targets.

Scientists at UCLA transformed HIV into a cancer seeking agent to infect cancer cells without provoking disease by beginning with a version of HIV that has the viral pieces that cause AIDS removed. The scientists then removed the HIV viral coat and re-dressed the outer suit in that of the sindbis virus which normally infects birds and insects. This alteration of HIV, using the sindbis coat made it so the virus attaches to P-glycoproteins rather than T cells, since P-glycoproteins are located on the surface of many cancer cells. Using this modified HIV; the scientist's used it on a mouse with cancer by injecting it into the blood stream and noted that the virus homed in on cancer cells. This experiment proves that a retrovirus can be utilized as a vector to attack cancer cells and modify genes within them. Although, using the using the sindbis virus coat would likely be ineffective in humans, as it is a bird and insect virus [219].

A retrovirus can separate the DNA of a host cell inserting RNA into each side of the DNA. As a result the genes noted in the DNA can be modified and apoptosis may be induced. There are certain retroviruses that I believe may be more beneficial to modify and use as vectors to stimulate or inhibit genes within cancer cells in order to stimulate and initiate apoptosis. I would like to see type C retroviruses experimented on for this type of gene therapy as these viruses alter the genes of the cells through reverse transcription and insertion of viral RNA that combines with host cell DNA.

HTLV2 Viral Diagram

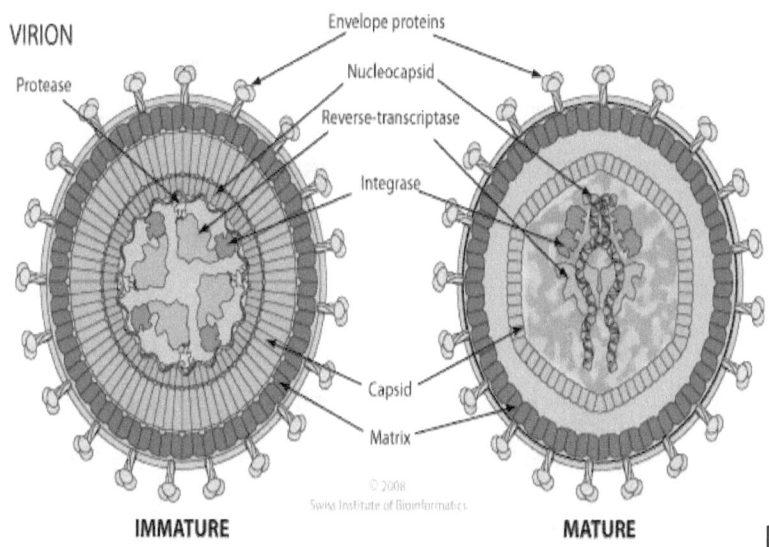

VIRION

Envelope proteins
Protease
Nucleocapsid
Reverse-transcriptase
Integrase
Capsid
Matrix
© 2008
Swiss Institute of Bioinformatics
IMMATURE
MATURE

[6p]

HTLV-2 Genome Diagram

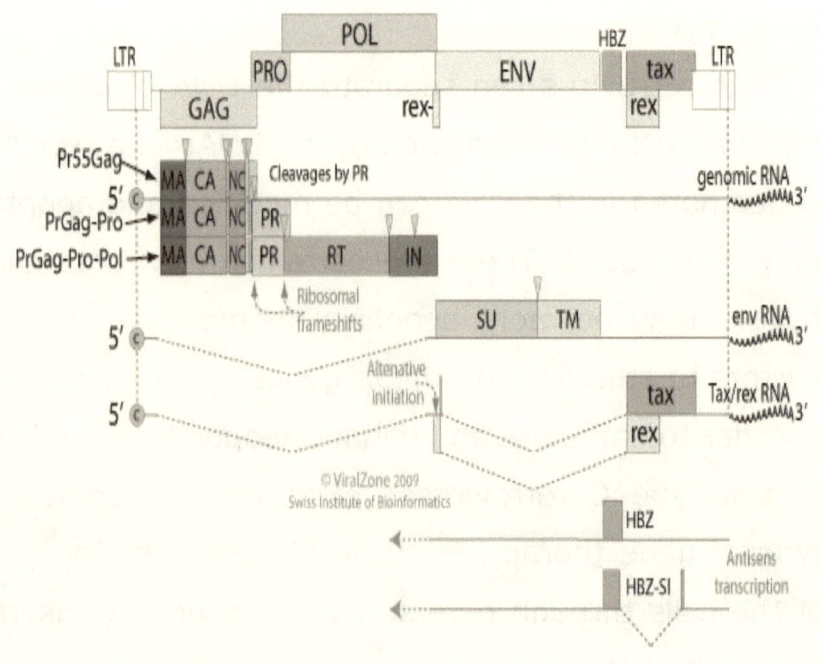

[6p]

The human t-cell lymphatic virus type two is a virus that can likely be used to genetically modify cancer cells to induce apoptosis. The human t-cell leukemia virus targets white blood cells, but that does not mean that it must infect only that cell type. The HTLV ENV gene can be modified to target other specific cell types. A genetically modified retrovirus can be utilized to modify the genetic code of targeted cancer cell DNA to induce apoptosis.

The HTLV2 GAG gene MA protein [p19] should be modified as it helps to target cancer cells. The GAG gene NC protein [p15] should be modified to alter host cell DNA after mRNA integration. The POL gene IN protein may be modified to integrate into a specific site on the host cell [220]. The HTLV-2 ENV gene SU protein [glycoprotein 46] should be modified to target various cell types as it binds to specific host cell receptors [221]. These genetic modifications should help to make HTLV2 into a vector that will induce cancer cells to begin apoptosis.

HTLV2 also contains genes that act from within the host cell nucleus after integration that include the TAX2 gene which activates transcription and the REX2 gene which is important for virus replication after transcription to export viral genes from the nucleus of the host cell [222, 223]. The REX gene proteins control how the virus reproduces in, and exports from virus infected cells. The REX gene would need to be modified in order to stop the virus from reproducing and spreading to other cells after infecting cancer cells to stimulate apoptosis. If all of these modifications could be made to the retrovirus it can be made into a cancer seeking agent that infects cancer cells to stimulate apoptosis. After apoptosis has been stimulated in cancer cells they will be removed from the body.

After a modified retrovirus is created to attack and stimulate apoptosis in a specific genetic type of cancer it may be reproduced to be used at a later point to treat the same cancer at a later time. Genetic cloning will likely be advantageous after a genetic bank of curative cancer cocktails are created as supplies will initially be limited. I hope scientists around the world who have more specialized knowledge regarding genetics, virology, microbiology, and oncology work together to help end humanity's struggle with cancer and cancer treatment to find a cure. While I have noted how my hypothesis may be applied to HTLV2, as that may or may not be the most advantageous retrovirus available to modify in an effort to stimulate apoptosis in cancer cells; or it may only be advantageous in particular cell types, as another retrovirus may be more beneficial to other cell types. Induced apoptosis can help to remove cancer cells from all parts of the body.

There would literally be many different cures for each cancer type since each cancer may have many genetic variations. I hope that experimental studies will help to successfully produce curative cancer treatments as a result of my scientific efforts. An initial curative cancer treatment cocktail that stimulates the proper apoptosis genes as the retrovirus GAG and POL genes are properly modified. This modified retrovirus may then be reproduced to simply modify the ENV gene within each specimen to target specific genetic cancer cell types that have the same general apoptotic genes needing to be stimulated. Of course the REX gene would need to be modified in order to make sure that the cancer seeking virus stops reproducing and is flushed from the body after all of a particular cancer cell type are destroyed.

A retrovirus ENV gene may be modified to target specific cancer cell types, while the GAG and POL genes may be modified to promote apoptosis in those specific cancer cell types via intrinsic apoptosis or extrinsic apoptosis, and the REX gene will need to be modified in order for the virus to stop attacking cells after cancer cells are destroyed being naturally flushed from the body. I hope that this may have initiated a curative cancer treatment campaign around the world. Hopefully suffering will be reduced as a result of my research.

Experiments

The general steps to modify the HTLV2 virus into a beneficial genetic vector to treat cancer should likely include five common steps. The primary reason that I am most fond of an HTLV vector and believe it to be the most advantageous retroviral vector is that it has the REX gene. The first step for using this genetic vector for any of the apoptosis experiments will be to modify the REX gene proteins in order to prevent the virus from reproducing and exporting from cells that the vector initially infects. If the REX gene proteins can be modified in this manner, we may have to use more of the vector to effectively treat the cancers, but it will be stopped from reproducing and spreading to other cells. The next two steps go hand in hand and depend upon each other. The second step will involve the modification of the POL gene IN protein in order for the virus to integrate into a specific site on the host cell chromosomes. The third step will involve the modification of the GAG gene NC protein in order to alter the host cell genes in a specific manner to promote apoptosis after integration. The fourth step involves the modification of the ENV gene SU protein so that this vector can be programmed to attack specific cancer cell lines, rather than just attacking T-cells as the virus does naturally. As long as we know what the specific genes are which identify a specific cell type then the ENV gene should be able to be modified and programmed to target that specific cell type. The fifth step may or may not be necessary depending upon tests, as the GAG gene MA protein may need to be modified in order to properly integrate the vector into the plasma membrane of the host cancer cells being treated. Of course the final two steps can truly only be carried out after a specific genetic line of cancer has been chosen as a target for this treatment. So, for each experiment I will note how the second and third steps may be applied to induce intrinsic and extrinsic apoptosis.

Experiment One: Intrinsic Apoptosis in Cancer Cells

The first step of this experiment would be to modify the REX gene so that it is modified to stop the virus from reproducing after it has infected a cell. The second step of this experiment will be making the virus attach to the BAK gene can be stimulated. In order to do this the HTLV2 POL gene IN protein will need to be modified in order to make the virus attach to the host cell chromosome on that specific point. The third step of this experiment will be making the virus alter the host cell chromosome in the proper manner to stimulate apoptosis in that cell. In this case we will need to modify the HTLV2 GAG gene NC protein in order to alter the host cell chromosome in the proper manner to stimulate apoptosis. This step may need a few experiments to find the proper stimulation.

The fourth step will involve choosing a specific cancer cell line in which to target by noting the specific genes that identify that cell type. In order to do this the HTLV2 ENV gene SU protein will need to be modified in order for the virus vector to attack a specific cell line that may have identifying genes or the cell line of interest may be targeted by noting the genetic mutations, many of which were noted throughout this book. The fifth step will be to modify the HTLV2 GAG gene MA protein in order for the vector to properly bond with the specific host cell plasma membrane. After modifications are made to the HTLV2 virus to turn it into a vector, intrinsic apoptosis will likely be possible as the p53 gene has been bypassed in the process.

Experiment Two: Extrinsic Apoptosis in Cancer Cells

For this experiment I would like to start by using two modified HTLV2 vectors; one to stimulate DR3 and another to stimulate death receptors 4 and 5. I like these two specific genes for stimulation to induce extrinsic apoptosis as DR3 and DR4 appear to be unaffected by the cancers noted in my study. While I am noting how the death receptor ligands noted above may be stimulated using my methods, this experiment should be applied in an effort to stimulate each death receptor ligands in order to activate coordinating cell death receptor, which in turn will cause extrinsic apoptosis and cell death. I am more confident that these experiments will likely render more advantageous results than experiment one, as this method is much simpler to initiate overall.

Just like the first experiment, the first step will be performed on the HTLV-2 REX gene proteins so they are modified to stop the virus from reproducing after it has infected a cell. We can then utilize this modified virus, that does not reproduce in and export from infected cells, for the next steps of the experiment to turn it into a beneficial vector.

The second step of this experiment will be to make one virus line to attach to the host cell chromosome 17 at the p13.1 point and another virus line to attach to the host cell chromosome 17 at the p13.3 point so the APO3L death receptor ligand gene can be stimulated. Of course another HTLV-2 virus that has been through the first two steps will need to be modified in order to attach to the host cell chromosome 3 at the q26 point so the APO2L death receptor ligand can be stimulated In order to do this the HTLV-2 POL gene IN protein will need to be modified in order to make the virus attach to the host cell chromosome on that specific point.

The third step of this experiment will be making each of the three viruses alter the host cell chromosome in the proper manner to stimulate apoptosis in that cell. In this case we will need to modify the HTLV-2 GAG gene NC protein [p15] in order to alter the host cell chromosome in the proper manner to stimulate apoptosis. This step may also need a few experiments on each modified virus to obtain the desired stimulation by each of them.

The fourth step will involve choosing a specific cancer cell line in which to target by noting the specific genes that identify that cell type. In order to do this the HTLV-2 ENV gene SU protein [glycoprotein 46] will need to be modified in order for the virus vector to attack a specific cell line that may have identifying genes or the cell line of interest may be targeted by noting the genetic mutations, many of which were noted throughout this book.

The fifth step will be to modify the HTLV-2 GAG gene MA protein in order for the vector to properly bond with the specific host cell plasma membrane.

After a specific genetic line of cancer is chosen to be targeted and modifications are made to the HTLV-2 virus to turn it into a vector, extrinsic apoptosis will likely be possible in any cell type targeted by this vector. As a curative cancer treatment process, I am much more confident in the extrinsic experiments as the cancers noted in my study show no mutations in the death receptor ligands, and minimal mutations in a few cancers of DR1 and DR5. I believe extrinsic apoptosis induced by a modified human T-cell leukemia virus is the key to curing cancer. And remember, just because the virus naturally attacks T-cells, does not mean that it must attack T-cells as the ENV gene can be modified!

Conclusion

There will be a curative cancer treatment cocktail for each different genetic line of cancer as my curative treatment process is applied all cancer types. There may be fewer total vectors created if target commonalities can be found for the various genetic lines of the same cancer types. I am confident that scientists around the world can work together, to be able to make humanity's problems caused as a result of cancer nothing more than a memory in less than a decade. I hope that scientist's experimenting with cancer cell apoptosis induced via retrovirus vector that has been reprogrammed to attack the cancer cells is very successful. I hope humanity can put our differences aside as we gain the motivation to end the pain caused as a result of cancer around the world since we now likely have an effective solution to end that pain.

As the medical scientists and nations of the earth come together and create genetically modified retroviruses that stimulate apoptosis in cancer cells by activating genes like DR3 and DR4. Then those viruses could have their ENV gene modified to attack any specific cancer cell/tissue types. Having genetic vectors modified as I have described, all of the cancer types noted in this book could be cured.

Author's Final Note

I hope to see the research from my book used to help end the pain caused by cancer. I am optimistic that my hypotheses will help to stimulate science to begin creating curative cancer treatments, rather than using many medical treatments that have harsh side effects. Hopefully I have sparked a worldwide curative cancer research campaign in which a curative cocktail vial for each genetic line of cancer is on the shelf as soon as possible. I have enjoyed composing this book in an effort to help medical science to take its first leap forward in a generation or more. I look forward to victims of cancer living with less pain as a result of utilizing a curative cancer treatment that will hopefully be on the doctors' shelves inside the next couple of years. I am a man that believes in tolerance and that relative world peace between nations and cultures on our earth is an accomplishable goal. I also hope that this book has positive scientific results and is a step in a direction of relative peace between cultures and nations as I have shared a free copy of "21st Century Cancer Treatment" and "A Curative Cancer Treatment", with the office of the leaders of every country on planet earth, Vice President Joe Biden, and the members of the United States Senate, as well as The Associated

Press. It is unfortunate that I live in a society that has been proven to be corrupt by money, as history has shown again and again that corporations and my government is more concerned about earning and continuing to gain monetary profits rather than help people to live better lives at a lower monetary cost. Class warfare is a BITCH!!! Other natural cancer remedies are noted at whale.to/cancer/therapies.html

References

1. American Society of Clinical Oncology. What is Cancer?. American Society of Clinical Oncology. "http://www.cancer.net/patient/Learning+About+Cancer/What +is+Cancer" [Updated June 13, 2008]. [Accessed December 2008].
2. National Cancer Institute. What is Cancer?. National Cancer Institute. "http://www.cancer.gov/cancertopics/what-is-cancer" [Accessed December 2008].
3. American Cancer Society. What is Cancer?. American Cancer Institute. "http://www.cancer.org/docroot/CRI/content/CRI_2_4_1x_What_Is_Cancer.asp?sitearea=" [Accessed December 2008].
4. Alberts B, Johnson A, Lewis J, Raff M, Roberts K, Walter P. "Molecular Biology of The Cell. 4th Edition". New York, NY: Garland Science, Taylor & Francis Group; 2002.
5. Unigene Organized View Of The Transcriptome. Baculoviral IAP Repeat-Containing 2 [BIRC2]. National Center for Biotechnology Information. "http://www.ncbi.nlm.nih.gov/UniGene/clust.cgi?UGID=2976843&TAXID=9606&SEARCH=BIRC2" [Accessed February 2011].
6. Online Mendelian Inheritance of Man, Johns Hopkins University. Baculoviral IAP Repeat Containing Protein 2; BIRC2. National Center for Biotechnology Information. "http://www.ncbi.nlm.nih.gov/omim/601712" [Updated December 3, 2010] [Accessed February 2011].
7. Unigene Organized View Of The Transcriptome. Baculoviral IAP Repeat-Containing 3 [BIRC3]. National Center for Biotechnology Information. "http://www.ncbi.nlm.nih.gov/UniGene/clust.cgi?UGID=147836&TAXID=9606&SEARCH=BIRC3" [Accessed February 2011].
8. Online Mendelian Inheritance of Man, Johns Hopkins University. Baculoviral IAP Repeat-Containing Protein 3; BIRC3. National Center for Biotechnology Information. "http://www.ncbi.nlm.nih.gov/omim/601721" [Updated December 3, 2010] [Accessed February 2011].
9. Unigene Organized View Of The Transcriptome. X-linked inhibitor of apoptosis [XIAP]. National Center for Biotechnology Information. "http://www.ncbi.nlm.nih.gov/UniGene/clust.cgi?UGID=1953

25&TAXID=9606&SEARCH=BIRC4" [Accessed February 2011].

10. Online Mendelian Inheritance of Man, Johns Hopkins University. Baculoviral IAP Repeat-Containing Protein 4; BIRC4. National Center for Biotechnology Information. "http://www.ncbi.nlm.nih.gov/omim/300079" [Updated September 15, 2009]. [Accessed February 2011].

11. Unigene Organized View Of The Transcriptome. Baculoviral IAP Repeat-Containing 5 [BIRC5]. National Center for Biotechnology Information. "http://www.ncbi.nlm.nih.gov/UniGene/clust.cgi?UGID=9052 74&TAXID=9606&SEARCH=BIRC5" [Accessed February 2011].

12. Online Mendelian Inheritance of Man, Johns Hopkins University. Baculoviral IAP Repeat-Containing Protein 5; BIRC5. National Center for Biotechnology Information. "http://www.ncbi.nlm.nih.gov/omim/603352" [Updated November 4, 2010]. [Accessed February 2011].

13. Unigene Organized View Of The Transcriptome. Apoptosis Inhibitor 5 [API5]. National Center for Biotechnology Information. "http://www.ncbi.nlm.nih.gov/UniGene/clust.cgi?UGID=2308 08&TAXID=9606&SEARCH=API5" [Accessed February 2011].

14. Online Mendelian Inheritance of Man, Johns Hopkins University. Apoptosis Inhibitor 5; API5. National Center for Biotechnology Information. "http://www.ncbi.nlm.nih.gov/omim/609774" [Updated December 12, 2005]. [Accessed February 2011].

15. Unigene Organized View Of The Transcriptome. Apoptosis, Caspase Activation Inhibitor [AVEN]. National Center for Biotechnology Information. "http://www.ncbi.nlm.nih.gov/UniGene/clust.cgi?UGID=1632 771&TAXID=9606&SEARCH=AVEN" [Accessed February 2011].

16. Online Mendelian Inheritance of Man, Johns Hopkins University. Cell Death Regulator AVEN. National Center for Biotechnology Information. "http://www.ncbi.nlm.nih.gov/omim/605265" [Updated September 14, 2000]. [Accessed February 2011].

17. Unigene Organized View Of The Transcriptome. B-Cell CLL/Lymphoma 2 [BCL2]. National Center for Biotechnology Information. "http://www.ncbi.nlm.nih.gov/UniGene/clust.cgi?UGID=1538 07&TAXID=9606&SEARCH=BCL2" [Accessed February 2011].

18. Online Mendelian Inheritance of Man, Johns Hopkins University. B-Cell CLL/Lymphoma 2; BCL2. National Center for Biotechnology Information. "http://www.ncbi.nlm.nih.gov/omim/151430" [Updated February 7, 2008]. [Accessed February 2011].

19. Unigene Organized View Of The Transcriptome.BCL2-Related Protein A1 [BCL2A1]. National Center for Biotechnology Information. "http://www.ncbi.nlm.nih.gov/UniGene/clust.cgi?UGID=1661 15&TAXID=9606&SEARCH=BCL2A1" [Accessed February 2011].

20. Online Mendelian Inheritance of Man, Johns Hopkins University. BCL2-Related Protein A1; BCL2A1. National Center for Biotechnology Information. "http://www.ncbi.nlm.nih.gov/omim/601056" [Updated January 25, 2007]. [Accessed February 2011].

21. Unigene Organized View Of The Transcriptome. BCL2-Like 1 [BCL2L1]. National Center for Biotechnology Information. "http://www.ncbi.nlm.nih.gov/UniGene/clust.cgi?UGID=9077 13&TAXID=9606&SEARCH=BCL-xL" [Accessed February 2011].

22. Online Mendelian Inheritance of Man, Johns Hopkins University. BCL2-Like 1; BCL2L1. National Center for Biotechnology Information. "http://www.ncbi.nlm.nih.gov/omim/600039" [Updated March 30, 2006]. [Accessed February 2011].

23. Unigene Organized View Of The Transcriptome. Baculoviral IAP Repeat-Containing 6 [BIRC6]. National Center for Biotechnology Information. "http://www.ncbi.nlm.nih.gov/UniGene/clust.cgi?UGID=1536 90&TAXID=9606&SEARCH=BIRC6" [Accessed February 2011].

24. Online Mendelian Inheritance of Man, Johns Hopkins University. Baculoviral IAP Repeat-Containing Protein 6; BIRC6. National Center for Biotechnology Information. "http://www.ncbi.nlm.nih.gov/omim/605638" [Updated February 2, 2009]. [Accessed February 2011].

25. Unigene Organized View Of The Transcriptome. Baculoviral IAP Repeat-Containing Protein 7 [BIRC7]. National Center for Biotechnology Information. "http://www.ncbi.nlm.nih.gov/UniGene/clust.cgi?UGID=1710 20&TAXID=9606&SEARCH=BIRC7" [Accessed February 2011].

26. Online Mendelian Inheritance of Man, Johns Hopkins University. Baculoviral IAP Repeat-Containing Protein 7;

BIRC7. National Center for Biotechnology Information. "http://www.ncbi.nlm.nih.gov/omim/605737" [Updated May 9, 2003]. [Accessed February 2011].

27. Unigene Organized View Of The Transcriptome. Cytokine Induced Apoptosis Inhibitor 1 [CIAPIN1]. National Center for Biotechnology Information. "http://www.ncbi.nlm.nih.gov/UniGene/clust.cgi?UGID=1315 75&TAXID=9606&SEARCH=CIAPIN1" [Accessed February 2011].

28. Online Mendelian Inheritance of Man, Johns Hopkins University. Cytokine-Induced Apoptosis Inhibitor 1; CIAPIN1. National Center for Biotechnology Information. "http://www.ncbi.nlm.nih.gov/omim/608943" [Updated November 2, 2004]. [Accessed February 2011].

29. Unigene Organized View Of The Transcriptome. Fas Apoptotic Inhibitory Molecule [FAIM]. National Center for Biotechnology Information. "http://www.ncbi.nlm.nih.gov/UniGene/clust.cgi?UGID=1583 58&TAXID=9606&SEARCH=FAIM" [Accessed February 2011].

30. Unigene Organized View Of The Transcriptome. Fas Apoptotic Inhibitory Molecule 2 [FAIM2]. National Center for Biotechnology Information. "http://www.ncbi.nlm.nih.gov/UniGene/clust.cgi?UGID=1782 913&TAXID=9606&SEARCH=FAIM2" [Accessed February 2011].

31. Online Mendelian Inheritance of Man, Johns Hopkins University. Fas Apoptotic Inhibitory Molecule 2; FAIM2. National Center for Biotechnology Information. "http://www.ncbi.nlm.nih.gov/omim/604306" [Updated November 19, 1999]. [Accessed February 2011].

32. Unigene Organized View Of The Transcriptome. Fas Apoptotic Inhibitory Molecule 3 [FAIM3]. National Center for Biotechnology Information. "http://www.ncbi.nlm.nih.gov/UniGene/clust.cgi?UGID=1385 57&TAXID=9606&SEARCH=FAIM3" [Accessed February 2011].

33. Unigene Organized View Of The Transcriptome. Myeloid Cell Leukemia Sequence 1 [BCL2-related] [MCL1]. National Center for Biotechnology Information. "http://www.ncbi.nlm.nih.gov/UniGene/clust.cgi?UGID=2139 380&TAXID=9606&SEARCH=MCL1" [Accessed February 2011].

34. Online Mendelian Inheritance of Man, Johns Hopkins University. Myeloid Cell Leukemia 1; MCL1. National Center for Biotechnology Information.

"http://www.ncbi.nlm.nih.gov/omim/159552"   [Updated
February 2, 2011]. [Accessed February 2011].

35. Unigene Organized View Of The Transcriptome. NLR Family,
Apoptosis Inhibitory Protein (NAIP). National Center for
Biotechnology   Information.
"http://www.ncbi.nlm.nih.gov/UniGene/clust.cgi?UGID=2723
818&TAXID=9606&SEARCH=NAIP"   [Accessed   February
2011].

36. Online Mendelian Inheritance of Man, Johns Hopkins
University. Baculoviral IAP Repeat-Containing Protein 1;
BIRC1. National Center for Biotechnology Information.
"http://www.ncbi.nlm.nih.gov/omim/600355"   [Updated
August 4, 2006]. [Accessed February 2011].

37. National Center for Biotechnology Information. Protein
Phosphatase 1, Regulatory [Inhibitor] Subunit 13 Like
[PPP1R13L]. Unigene Organized View Of The Transcriptome.
"http://www.ncbi.nlm.nih.gov/UniGene/clust.cgi?UGID=6820
96&TAXID=9606&SEARCH=PPP1R13L" [Accessed   February
2011].

38. Online Mendelian Inheritance of Man, Johns Hopkins
University. Protein Phosphatase 1, Regulatory Subunit 13-
Like; PPP1R13L. National Center for Biotechnology
Information.   "http://www.ncbi.nlm.nih.gov/omim/607463"
[Updated may 27, 2008]. [Accessed February 2011].

39. Unigene Organized View Of The Transcriptome. Protein
Phosphatase 1, Regulatory [Inhibitor] Subunit 15A
[PPP1R15A]. National Center for Biotechnology Information.
"http://www.ncbi.nlm.nih.gov/UniGene/clust.cgi?UGID=2138
487&TAXID=9606&SEARCH=PPP1R15A" [Accessed February
2011].

40. Online Mendelian Inheritance of Man, Johns Hopkins
University. Protein Phosphatase 1, Regulatory Subunit 15A;
PPP1R15A. National Center for Biotechnology Information.
"http://www.ncbi.nlm.nih.gov/omim/611048" [Updated May
21, 2007]. [Accessed February 2011].

41. Unigene Organized View Of The Transcriptome. Synovial
Apoptosis Inhibitor 1 [SYVN1]. National Center for
Biotechnology   Information.
"http://www.ncbi.nlm.nih.gov/UniGene/clust.cgi?UGID=3547
754&TAXID=9606&SEARCH=SYVN1" [Accessed July 2009].

42. Online Mendelian Inheritance of Man, Johns Hopkins
University. Synovial Apoptosis Inhibitor 1; SYVN1. National
Center   for   Biotechnology   Information.
"http://www.ncbi.nlm.nih.gov/omim/608046"   [Updated
November 10, 2008]. [Accessed February 2011].

43. Unigene Organized View Of The Transcriptome. TP53 Regulated Inhibitor of Apoptosis 1 [TRIAP1]. National Center for Biotechnology Information. "http://www.ncbi.nlm.nih.gov/UniGene/clust.cgi?UGID=139315&TAXID=9606&SEARCH=TRIAP1" [Accessed February 2011].

44. Online Mendelian Inheritance of Man, Johns Hopkins University. Apoptosis Regulator BCLG. National Center for Biotechnology Information. "http://www.ncbi.nlm.nih.gov/omim/606126" [Updated August 29, 2002]. [Accessed February 2011].

45. Unigene Organized View Of The Transcriptome. BCL2-like 2 [BCL2L2]. National Center for Biotechnology Information. "http://www.ncbi.nlm.nih.gov/UniGene/clust.cgi?UGID=221469&TAXID=9606&SEARCH=BCL2L2" [Accessed February 2011].

46. Online Mendelian Inheritance of Man, Johns Hopkins University. BCL2-Like2; BCL2L2. National Center for Biotechnology Information. "http://www.ncbi.nlm.nih.gov/omim/601931" [Updated February 27, 1998]. [Accessed February 2011].

47. Unigene Organized View Of The Transcriptome. Bifunctional Apoptosis Regulator [BFAR]. National Center for Biotechnology Information. "http://www.ncbi.nlm.nih.gov/UniGene/clust.cgi?UGID=230593&TAXID=9606&SEARCH=BFAR" [Accessed February 2011].

48. Unigene Organized View Of The Transcriptome. Cell Division Cycle and Apoptosis Regulator 1 [CCAR1]. National Center for Biotechnology Information. "http://www.ncbi.nlm.nih.gov/UniGene/clust.cgi?UGID=137878&TAXID=9606&SEARCH=CCAR1" [Accessed February 2011].

49. Online Mendelian Inheritance of Man, Johns Hopkins University. Cell Division Cycle and Apoptosis Regulator 1; CCAR1. National Center for Biotechnology Information. "http://www.ncbi.nlm.nih.gov/omim/612569" [Updated February 2, 2009]. [Accessed February 2011].

50. Unigene Organized View Of The Transcriptome. CASP8 and FADD-Like Apoptosis Regulator [CFLAR]. National Center for Biotechnology Information. "http://www.ncbi.nlm.nih.gov/UniGene/clust.cgi?UGID=208587&TAXID=9606&SEARCH=CFLAR" [Accessed February 2011].

51. Online Mendelian Inheritance of Man, Johns Hopkins University. CASP8- and FADD-Like Apoptosis Regulator;

CFLAR. National Center for Biotechnology Information. "http://www.ncbi.nlm.nih.gov/omim/603599" [Updated May 21, 2009]. [Accessed February 2011].

52. Unigene Organized View Of The Transcriptome. PRKC, Apoptosis, WT1, Regulator [PAWR]. National Center for Biotechnology Information. "http://www.ncbi.nlm.nih.gov/UniGene/clust.cgi?UGID=2224 318&TAXID=9606&SEARCH=PAWR" [Accessed February 2011].

53. Online Mendelian Inheritance of Man, Johns Hopkins University. PRKC, Apoptosis, WT1, Regulator; PAWR. National Center for Biotechnology Information. "http://www.ncbi.nlm.nih.gov/omim/601936" [Updated August 5, 2005]. [Accessed February 2011].

54. Unigene Organized View Of The Transcriptome. Tumor Protein p53 [TP53]. National Center for Biotechnology Information. "http://www.ncbi.nlm.nih.gov/UniGene/clust.cgi?UGID=2723 799&TAXID=9606&SEARCH=TP53" [Accessed February 2011].

55. Online Mendelian Inheritance of Man, Johns Hopkins University. Tumor Protein p53; TP53. National Center for Biotechnology Information. "http://www.ncbi.nlm.nih.gov/omim/191170" [Updated February 7, 2011]. [Accessed February 2011].

56. Unigene Organized View Of The Transcriptome. Modulator of Apoptosis 1 [MOAP1]. National Center for Biotechnology Information. "http://www.ncbi.nlm.nih.gov/UniGene/clust.cgi?UGID=1348 94&TAXID=9606&SEARCH=MOAP1" [Accessed February 2011].

57. Online Mendelian Inheritance of Man, Johns Hopkins University. Modulator of Apoptosis 1; MOAP1. National Center for Biotechnology Information. "http://www.ncbi.nlm.nih.gov/omim/609485" [Updated July 21, 2005]. [Accessed February 2011].

58. Unigene Organized View Of The Transcriptome. BCL2 Binding Component 3 [BBC3]. National Center for Biotechnology Information. "http://www.ncbi.nlm.nih.gov/UniGene/clust.cgi?UGID=6821 79&TAXID=9606&SEARCH=BBC3" [Accessed February 2011].

59. Online Mendelian Inheritance of Man, Johns Hopkins University. BCL2-Binding Component 3; BBC3. National Center for Biotechnology Information. "http://www.ncbi.nlm.nih.gov/omim/605854" [Updated December 28, 2010]. [Accessed February 2011].

60. Unigene Organized View Of The Transcriptome. BCL2-Like 10 [Apoptosis Facilitator] [BCL2L10]. National Center for Biotechnology Information. "http://www.ncbi.nlm.nih.gov/UniGene/clust.cgi?UGID=1767 85&TAXID=9606&SEARCH=BCL2L10" [Accessed February 2011].

61. Online Mendelian Inheritance of Man, Johns Hopkins University. BCL2-Like 10; BCL2L10. National Center for Biotechnology Information. "http://www.ncbi.nlm.nih.gov/omim/606910" [Updated June 7, 2002]. [Accessed February 2011].

62. Unigene Organized View Of The Transcriptome. BCL2-Like 11 [Apoptosis Facilitator] [BCL2L11]. National Center for Biotechnology Information. "http://www.ncbi.nlm.nih.gov/UniGene/clust.cgi?UGID=6848 17&TAXID=9606&SEARCH=BCL2L11" [Accessed February 2011].

63. Online Mendelian Inheritance of Man, Johns Hopkins University. BCL2-Like 11; BCL2L11. National Center for Biotechnology Information. "http://www.ncbi.nlm.nih.gov/omim/603827" [Updated December 28, 2010]. [Accessed February 2011].

64. Unigene Organized View Of The Transcriptome. BCL2-Like 13 [Apoptosis Facilitator] [BCL2L13]. National Center for Biotechnology Information. "http://www.ncbi.nlm.nih.gov/UniGene/clust.cgi?UGID=2138 566&TAXID=9606&SEARCH=BCL2L13" [Accessed February 2011].

65. Unigene Organized View Of The Transcriptome. BCL2-Like 14 [Apoptosis Facilitator] [BCL2L14]. National Center for Biotechnology Information. "http://www.ncbi.nlm.nih.gov/UniGene/clust.cgi?UGID=1640 35&TAXID=9606&SEARCH=BCL2L14" [Accessed February 2011].

66. Unigene Organized View Of The Transcriptome. Apoptosis-Inducing Factor, Mitochondrion Associated, 1 [AIFM1]. National Center for Biotechnology Information. "http://www.ncbi.nlm.nih.gov/UniGene/clust.cgi?UGID=2253 63&TAXID=9606&SEARCH=AIFM1" [Accessed February 2011].

67. Online Mendelian Inheritance of Man, Johns Hopkins University. Apoptosis-Inducing Factor, Mitochondrion Associated, 1; AIFM1. National Center for Biotechnology Information. "http://www.ncbi.nlm.nih.gov/omim/300169" [Updated April 30, 2010]. [Accessed February 2011].

68. Unigene Organized View Of The Transcriptome. Apoptosis-Inducing Factor, Mitochondrion Associated, 2 [AIFM2]. National Center for Biotechnology Information. "http://www.ncbi.nlm.nih.gov/UniGene/clust.cgi?UGID=2724 695&TAXID=9606&SEARCH=AIFM2" [Accessed February 2011].

69. Online Mendelian Inheritance of Man, Johns Hopkins University. Apoptosis-Inducing Factor, Mitochondrion Associated, 2; AIFM2. National Center for Biotechnology Information. "http://www.ncbi.nlm.nih.gov/omim/605159" [Updated July 19, 2000]. [Accessed February 2011].

70. Unigene Organized View Of The Transcriptome. Cortistatin [CORT]. National Center for Biotechnology Information. "http://www.ncbi.nlm.nih.gov/UniGene/clust.cgi?UGID=2219 89&TAXID=9606&SEARCH=CORT" [Accessed February 2011].

71. Online Mendelian Inheritance of Man, Johns Hopkins University. Apoptosis-Inducing, TAF9-Like Domain 1; APITD1. National Center for Biotechnology Information. "http://www.ncbi.nlm.nih.gov/omim/609130" [Updated October 8, 2007]. [Accessed February 2011].

72. Unigene Organized View Of The Transcriptome. BCL2-Associated Agonist of Cell Death [BAD]. National Center for Biotechnology Information. "http://www.ncbi.nlm.nih.gov/UniGene/clust.cgi?UGID=1985 54&TAXID=9606&SEARCH=BAD" [Accessed February 2011].

73. Online Mendelian Inheritance of Man, Johns Hopkins University. BCL2 Antagonist of Cell Death; BAD. National Center for Biotechnology Information. "http://www.ncbi.nlm.nih.gov/omim/603167" [Updated May 10, 2010]. [Accessed February 2011].

74. Unigene Organized View Of The Transcriptome. BCL2-Antagonist/Killer 1 [BAK1]. National Center for Biotechnology Information. "http://www.ncbi.nlm.nih.gov/UniGene/clust.cgi?UGID=7002 98&TAXID=9606&SEARCH=BAK1" [Accessed February 2011].

75. Online Mendelian Inheritance of Man, Johns Hopkins University. BCL2 Antagonist Killer 1; BAK1. National Center for Biotechnology Information. "http://www.ncbi.nlm.nih.gov/omim/600516" [Updated December 28, 2010]. [Accessed February 2011].

76. Unigene Organized View Of The Transcriptome. BCL2-Associated X Protein [BAX]. National Center for Biotechnology Information. "http://www.ncbi.nlm.nih.gov/UniGene/clust.cgi?UGID=2093 302&TAXID=9606&SEARCH=BAX" [Accessed February 2011].

77.   Online Mendelian Inheritance of Man, Johns Hopkins University. BCL2-Associated X Protein; BAX. National Center for Biotechnology Information. "http://www.ncbi.nlm.nih.gov/omim/600040" [Updated December 28, 2010]. [Accessed February 2011].

78. Unigene Organized View Of The Transcriptome. BCL2-Like 15 [BCL2L15]. National Center for Biotechnology Information. "http://www.ncbi.nlm.nih.gov/UniGene/clust.cgi?UGID=1464 12&TAXID=9606&SEARCH=BCL2L15" [Accessed February 2011].

79.   Unigene Organized View Of The Transcriptome. BH3 Interacting Domain Death Agonist [BID]. National Center for Biotechnology Information. "http://www.ncbi.nlm.nih.gov/UniGene/clust.cgi?UGID=2060 065&TAXID=9606&SEARCH=BID" [Accessed February 2011].

80.   Online Mendelian Inheritance of Man, Johns Hopkins University. BH3 Interacting Domain Death Agonist; BID. National Center for Biotechnology Information. "http://www.ncbi.nlm.nih.gov/omim/601997" [Updated December 28, 2010]. [Accessed February 2011].

81.   Unigene Organized View Of The Transcriptome. BCL2-Interacting Killer [Apoptosis-Inducing] [BIK]. National Center for Biotechnology Information. "http://www.ncbi.nlm.nih.gov/UniGene/clust.cgi?UGID=6902 14&TAXID=9606&SEARCH=BIK" [Accessed February 2011].

82.   Online Mendelian Inheritance of Man, Johns Hopkins University. BCL2-Interacting Killer; BIK. National Center for Biotechnology Information. "http://www.ncbi.nlm.nih.gov/omim/603392" [Updated February 9, 2001]. [Accessed February 2011].

83.   Unigene Organized View Of The Transcriptome. BCL2L11. National Center for Biotechnology Information. "http://www.ncbi.nlm.nih.gov/UniGene/clust.cgi?UGID=6848 17&TAXID=9606&SEARCH=BCL2L11" [Accessed February 2011].

84.   Online Mendelian Inheritance of Man, Johns Hopkins University. BCL2-LIKE 11; BCL2L11. National Center for Biotechnology Information. "http://www.ncbi.nlm.nih.gov/omim/603827" [Updated December 28, 2010]. [Accessed February 2011].

85.   Unigene Organized View Of The Transcriptome. BCL2 Modifying Factor [BMF]. National Center for Biotechnology Information. "http://www.ncbi.nlm.nih.gov/UniGene/clust.cgi?UGID=2060 115&TAXID=9606&SEARCH=BMF" [Accessed February 2011].

86. Online Mendelian Inheritance of Man, Johns Hopkins University. BCL2 Modifying Factor; BMF. National Center for Biotechnology Information. "http://www.ncbi.nlm.nih.gov/omim/606266" [Updated September 17, 2001]. [Accessed February 2011].

87. Unigene Organized View Of The Transcriptome. BCL2-Related Ovarian Killer [BOK]. National Center for Biotechnology Information. "http://www.ncbi.nlm.nih.gov/UniGene/clust.cgi?UGID=1798 31&TAXID=9606&SEARCH=BOK" [Accessed February 2011].

88. Online Mendelian Inheritance of Man, Johns Hopkins University. BCL2-Related Ovarian Killer; BOK. National Center for Biotechnology Information. "http://www.ncbi.nlm.nih.gov/omim/605404" [Updated November 14, 2000]. [Accessed February 2011].

89. Unigene Organized View Of The Transcriptome. Cell Death-Inducing DFFA Like Effector B [CIDEB]. National Center for Biotechnology Information. "http://www.ncbi.nlm.nih.gov/UniGene/clust.cgi?UGID=2223 949&TAXID=9606&SEARCH=CIDEB" [Accessed February 2011].

90. Online Mendelian Inheritance of Man, Johns Hopkins University. Cell Death-Inducing DFFA-Like Effector B. National Center for Biotechnology Information. "http://www.ncbi.nlm.nih.gov/omim/604441" [Updated January 19, 2000]. [Accessed February 2011].

91. Unigene Organized View Of The Transcriptome. Harakiri, BCL2 Interacting Protein [Contains Only BH3 Domain] [HRK]. National Center for Biotechnology Information. "http://www.ncbi.nlm.nih.gov/UniGene/clust.cgi?UGID=1410 36&TAXID=9606&SEARCH=HRK" [Accessed February 2011].

92. Online Mendelian Inheritance of Man, Johns Hopkins University. Harakiri; HRK. National Center for Biotechnology Information. "http://www.ncbi.nlm.nih.gov/omim/603447" [Updated June 11, 2007]. [Accessed February 2011].

93. Unigene Organized View Of The Transcriptome. Mitochondrial Ribosomal Protein S30 [MRPS30]. National Center for Biotechnology Information. "http://www.ncbi.nlm.nih.gov/UniGene/clust.cgi?UGID=1466 79&TAXID=9606&SEARCH=MRPS30" [Accessed February 2011].

94. Online Mendelian Inheritance of Man, Johns Hopkins University. Mitochondrial Ribosomal Protein S30; MRPS30. National Center for Biotechnology Information.

"http://www.ncbi.nlm.nih.gov/omim/611991" [Updated April 21, 2008]. [Accessed February 2011].

95. Unigene Organized View Of The Transcriptome. Nuclear Apoptosis Inducing Factor [NAIF1]. National Center for Biotechnology Information. "http://www.ncbi.nlm.nih.gov/UniGene/clust.cgi?UGID=2002 80&TAXID=9606&SEARCH=NAIF1" [Accessed February 2011].

96. Online Mendelian Inheritance of Man, Johns Hopkins University. Chromosome 9 Open Reading Frame 90; C9ORF90. National Center for Biotechnology Information. "http://www.ncbi.nlm.nih.gov/omim/610673" [Updated December 28, 2006]. [Accessed February 2011].

97. Unigene Organized View Of The Transcriptome. Phorbol-12-Myristate-13-Acetate-Induced Protein 1 [PMAIP1]. National Center for Biotechnology Information. "http://www.ncbi.nlm.nih.gov/UniGene/clust.cgi?UGID=1306 71&TAXID=9606&SEARCH=PMAIP1" [Accessed February 2011].

98. Online Mendelian Inheritance of Man, Johns Hopkins University. Phorbol-12-Myristate-13-Acetate-Induced Protein 1; PMAIP1. National Center for Biotechnology Information. "http://www.ncbi.nlm.nih.gov/omim/604959" [Updated October 26, 2005]. [Accessed February 2011].

99. Unigene Organized View Of The Transcriptome. Programmed Cell Death 1 [PDCD1]. National Center for Biotechnology Information. "http://www.ncbi.nlm.nih.gov/UniGene/clust.cgi?UGID=1552 58&TAXID=9606&SEARCH=PDCD1" [Accessed February 2011].

100. Online Mendelian Inheritance of Man, Johns Hopkins University. Programmed Cell Death 1; PDCD1. National Center for Biotechnology Information. "http://www.ncbi.nlm.nih.gov/omim/600244" [Updated December 12, 2010]. [Accessed February 2011].

101. Unigene Organized View Of The Transcriptome. Programmed Cell Death 2 [PDCD2]. National Center for Biotechnology Information. "http://www.ncbi.nlm.nih.gov/UniGene/clust.cgi?UGID=1972 14&TAXID=9606&SEARCH=PDCD2" [Accessed February 2011].

102. Online Mendelian Inheritance of Man, Johns Hopkins University. Programmed Cell Death 2; PDCD2. National Center for Biotechnology Information.

"http://www.ncbi.nlm.nih.gov/omim/600866" [Updated June 11, 2007]. [Accessed February 2011].

103. Unigene Organized View Of The Transcriptome. Programmed Cell Death 4 [Neoplastic Transformation Inhibitor] [PDCD4]. National Center for Biotechnology Information. "http://www.ncbi.nlm.nih.gov/UniGene/clust.cgi?UGID=3323 697&TAXID=9606&SEARCH=PDCD4" [Accessed February 2011].

104. Online Mendelian Inheritance of Man, Johns Hopkins University. Programmed Cell Death 4; PDCD4. National Center for Biotechnology Information. "http://www.ncbi.nlm.nih.gov/omim/608610" [Updated December 12, 2010]. [Accessed February 2011].

105. Unigene Organized View Of The Transcriptome. Programmed Cell Death 5 [PDCD5]. National Center for Biotechnology Information. "http://www.ncbi.nlm.nih.gov/UniGene/clust.cgi?UGID=2388 68&TAXID=9606&SEARCH=PDCD5" [Accessed February 2011].

106. Online Mendelian Inheritance of Man, Johns Hopkins University. Programmed Cell Death 5; PDCD5. National Center for Biotechnology Information. "http://www.ncbi.nlm.nih.gov/omim/604583" [Updated February 19, 2000]. [Accessed February 2011].

107. Unigene Organized View Of The Transcriptome. Programmed Cell Death 6 [PDCD6]. National Center for Biotechnology Information. "http://www.ncbi.nlm.nih.gov/UniGene/clust.cgi?UGID=1379 96&TAXID=9606&SEARCH=PDCD6" [Accessed February 2011].

108. Online Mendelian Inheritance of Man, Johns Hopkins University. Programmed Cell Death 6; PDCD6. National Center for Biotechnology Information. "http://www.ncbi.nlm.nih.gov/omim/601057" [Updated April 7, 2006]. [Accessed February 2011].

109. Unigene Organized View Of The Transcriptome. Programmed Cell Death 6 Interacting Protein [PDCD6IP]. National Center for Biotechnology Information. "http://www.ncbi.nlm.nih.gov/UniGene/clust.cgi?UGID=6910 55&TAXID=9606&SEARCH=PDCD6IP" [Accessed February 2011].

110. Online Mendelian Inheritance of Man, Johns Hopkins University. Programmed Cell Death 6 Interacting Protein; PDCD6IP. National Center for Biotechnology Information.

"http://www.ncbi.nlm.nih.gov/omim/608074" [Updated July 24, 2007]. [Accessed February 2011].

111. Unigene Organized View Of The Transcriptome. Programmed Cell Death 7 [PDCD7]. National Center for Biotechnology Information. "http://www.ncbi.nlm.nih.gov/UniGene/clust.cgi?UGID=6737 55&TAXID=9606&SEARCH=PDCD7" [Accessed February 2011].

112. Online Mendelian Inheritance of Man, Johns Hopkins University. Programmed Cell Death 7; PDCD7. National Center for Biotechnology Information. "http://www.ncbi.nlm.nih.gov/omim/608138" [Updated September 30, 2003]. [Accessed February 2011].

113. Unigene Organized View Of The Transcriptome. Programmed Cell Death 10 [PDCD10]. National Center for Biotechnology Information. "http://www.ncbi.nlm.nih.gov/UniGene/clust.cgi?UGID=6933 09&TAXID=9606&SEARCH=PDCD10" [Accessed February 2011].

114. Online Mendelian Inheritance of Man, Johns Hopkins University. Programmed Cell Death 10; PDCD10. National Center for Biotechnology Information. "http://www.ncbi.nlm.nih.gov/omim/609118" [Updated August 12, 2009]. [Accessed February 2011].

115. Unigene Organized View Of The Transcriptome. Programmed Cell Death 11 [PDCD11]. National Center for Biotechnology Information. "http://www.ncbi.nlm.nih.gov/UniGene/clust.cgi?UGID=1677 76&TAXID=9606&SEARCH=PDCD11" [Accessed February 2011].

116. Online Mendelian Inheritance of Man, Johns Hopkins University. Programmed Cell Death 11; PDCD11. National Center for Biotechnology Information. "http://www.ncbi.nlm.nih.gov/omim/612333" [Updated September 29, 2008]. [Accessed February 2011].

117. Unigene Organized View Of The Transcriptome. PERP, TP53 Apoptosis Effector [PERP]. National Center for Biotechnology Information. "http://www.ncbi.nlm.nih.gov/UniGene/clust.cgi?UGID=1623 75&TAXID=9606&SEARCH=PERP" [Accessed February 2011].

118. Online Mendelian Inheritance of Man, Johns Hopkins University. p53 Effector Related To PMP22; PERP. National Center for Biotechnology Information. "http://www.ncbi.nlm.nih.gov/omim/609301" [Updated April 6, 2005]. [Accessed February 2011].

119. Unigene Organized View Of The Transcriptome. SIVA1, Apoptosis-Inducing Factor [SIVA1]. National Center for Biotechnology Information. "http://www.ncbi.nlm.nih.gov/UniGene/clust.cgi?UGID=1441 44&TAXID=9606&SEARCH=SIVA1" [Accessed February 2011].

120. Online Mendelian Inheritance of Man, Johns Hopkins University. SIVA Apoptosis-Inducing Factor 1; SIVA1. National Center for Biotechnology Information. "http://www.ncbi.nlm.nih.gov/omim/605567" [Updated November 11, 2005]. [Accessed February 2011].

121. Unigene Organized View Of The Transcriptome. Tumor Protein p53 Regulated Apoptosis Inducing Protein 1 [TP53AIP1]. National Center for Biotechnology Information. "http://www.ncbi.nlm.nih.gov/UniGene/clust.cgi?UGID=1559 17&TAXID=9606&SEARCH=TP53AIP1" [Accessed February 2011].

122. Online Mendelian Inheritance of Man, Johns Hopkins University. P53-Regulated Apoptosis-Inducing Protein1. National Center for Biotechnology Information. "http://www.ncbi.nlm.nih.gov/omim/605426" [Updated November 28, 2000]. [Accessed February 2011].

123. Unigene Organized View Of The Transcriptome. Fas Ligand [TNF Superfamily, Member 6] [FASLG]. National Center for Biotechnology Information. "http://www.ncbi.nlm.nih.gov/UniGene/clust.cgi?UGID=1311 33&TAXID=9606&SEARCH=TNFSF6" [Accessed February 2011].

124. Online Mendelian Inheritance of Man, Johns Hopkins University. Tumor Necrosis Factor Ligand Superfamily, Member 6; TNFSF6. National Center for Biotechnology Information. "http://www.ncbi.nlm.nih.gov/omim/134638" [Updated November 16, 2009]. [Accessed February 2011].

125. Unigene Organized View Of The Transcriptome. Lymphotoxin Alpha [TNF, Superfamily, Member 1] [LTA]. National Center for Biotechnology Information. "http://www.ncbi.nlm.nih.gov/UniGene/clust.cgi?UGID=1306 54&TAXID=9606&SEARCH=TNF-BETA" [Accessed February 2011].

126. Online Mendelian Inheritance of Man, Johns Hopkins University. Lymphotoxin-Alpha; LTA. National Center for Biotechnology Information. "http://www.ncbi.nlm.nih.gov/omim/153440" [Updated June 4, 2007]. [Accessed February 2011].

127. Unigene Organized View Of The Transcriptome. Tumor Necrosis Factor [Ligand] Superfamily, Member 13 [TNFSF13]. National Center for Biotechnology Information. "http://www.ncbi.nlm.nih.gov/UniGene/clust.cgi?UGID=1381 54&TAXID=9606&SEARCH=TNFSF12" [Accessed February 2011].
128. Online Mendelian Inheritance of Man, Johns Hopkins University. Tumor Necrosis Factor Ligand Superfamily, Member 12; TNFSF12. National Center for Biotechnology Information. "http://www.ncbi.nlm.nih.gov/omim/602695" [Updated May 10, 2007] [Accessed February 2011].
129. Unigene Organized View Of The Transcriptome. Tumor Necrosis Factor [Ligand] Superfamily, Member 10 [TNFSF10]. National Center for Biotechnology Information. "http://www.ncbi.nlm.nih.gov/UniGene/clust.cgi?UGID=6934 34&TAXID=9606&SEARCH=TNFSF10" [Accessed February 2011].
130. Online Mendelian Inheritance of Man, Johns Hopkins University. Tumor Necrosis Factor [Ligand] Superfamily, Member 10; TNFSF10. National Center for Biotechnology Information. "http://www.ncbi.nlm.nih.gov/omim/603598" [Updated May 10, 2010]. [Accessed February 2011].
131. Unigene Organized View Of The Transcriptome. CD274 Molecule [CD274]. National Center for Biotechnology Information. "http://www.ncbi.nlm.nih.gov/UniGene/clust.cgi?UGID=9127 36&TAXID=9606&SEARCH=PDCD1LG1" [Accessed February 2011].
132. Online Mendelian Inheritance of Man, Johns Hopkins University. Programmed Cell Death 1 Ligand 1; PDCDLG1. National Center for Biotechnology Information. "http://www.ncbi.nlm.nih.gov/omim/605402" [Updated June 7, 2010]. [Accessed February 2011].
133. Unigene Organized View Of The Transcriptome. Programmed Cell Death 1 Ligand 2 [PDCD1LG2]. National Center for Biotechnology Information. "http://www.ncbi.nlm.nih.gov/UniGene/clust.cgi?UGID=1272 177&TAXID=9606&SEARCH=PDCD1LG2" [Accessed February 2011].
134. Online Mendelian Inheritance of Man, Johns Hopkins University. Programmed Cell Death 1 Ligand 2; PDCD1LG2. National Center for Biotechnology Information. "http://www.ncbi.nlm.nih.gov/omim/605723" [Updated October 3, 2006]. [Accessed February 2011].

135. Unigene Organized View Of The Transcriptome. Fas [TNF Receptor Superfamily, Member 6] [FAS]. National Center for Biotechnology Information. "http://www.ncbi.nlm.nih.gov/UniGene/clust.cgi?UGID=168360&TAXID=9606&SEARCH=FAS" [Accessed February 2011].

136. Online Mendelian Inheritance of Man, Johns Hopkins University. Tumor Necrosis Factor Receptor Superfamily, Member 6; TNFRSF6. National Center for Biotechnology Information. "http://www.ncbi.nlm.nih.gov/omim/134637" [Updated June, 30, 2010]. [Accessed February 2011].

137. Unigene Organized View Of The Transcriptome. Tumor Necrosis Factor Receptor Superfamily, Member 1A [TNFRSF1A]. National Center for Biotechnology Information. "http://www.ncbi.nlm.nih.gov/UniGene/clust.cgi?UGID=175356&TAXID=9606&SEARCH=TNFRSF1a" [Accessed February 2011].

138. Online Mendelian Inheritance of Man, Johns Hopkins University. Tumor Necrosis Factor Receptor Superfamily, Member 1A; TNFRSF1A. National Center for Biotechnology Information. "http://www.ncbi.nlm.nih.gov/omim/191190" [Updated September 9, 2009]. [Accessed February 2011].

139. Unigene Organized View Of The Transcriptome. Tumor Necrosis Factor Receptor Superfamily, Member 25 [TNFRSF25]. National Center for Biotechnology Information. "http://www.ncbi.nlm.nih.gov/UniGene/clust.cgi?UGID=677688&TAXID=9606&SEARCH=TNFRSF25" [Accessed February 2011].

140. Online Mendelian Inheritance of Man, Johns Hopkins University. Tumor Necrosis Factor Receptor Superfamily, Member 25; TNFRSF25. National Center for Biotechnology Information. "http://www.ncbi.nlm.nih.gov/omim/603366" [Updated January 20, 2010]. [Accessed February 2011].

141. Unigene Organized View Of The Transcriptome. Tumor Necrosis Factor Receptor Superfamily, Member 10A [TNFRSF10A]. National Center for Biotechnology Information. "http://www.ncbi.nlm.nih.gov/UniGene/clust.cgi?UGID=2060845&TAXID=9606&SEARCH=TNFRSF10" [Accessed February 2011].

142. Online Mendelian Inheritance of Man, Johns Hopkins University. Tumor Necrosis Factor Receptor Superfamily, Member 10A; TNFRSF10A. National Center for Biotechnology Information. "http://www.ncbi.nlm.nih.gov/omim/603611" [Updated March 13, 2008]. [Accessed February 2011].

143. Unigene Organized View Of The Transcriptome. Tumor Necrosis Factor Receptor Superfamily, Member 10B

[TNFRSF10B]. National Center for Biotechnology Information. "http://www.ncbi.nlm.nih.gov/UniGene/clust.cgi?UGID=9122 03&TAXID=9606&SEARCH=TNFRSF10B" [Accessed February 2011].

144. Online Mendelian Inheritance of Man, Johns Hopkins University. Tumor Necrosis Factor Receptor Superfamily, Member 10B; TNFRSF10B. National Center for Biotechnology Information. "http://www.ncbi.nlm.nih.gov/omim/603612" [Updated April 29, 2009]. [Accessed February 2011].

145. Unigene Organized View Of The Transcriptome. Apoptosis Antagonizing Transcription Factor [AATF]. National Center for Biotechnology Information. "http://www.ncbi.nlm.nih.gov/UniGene/clust.cgi?UGID=1614 56&TAXID=9606&SEARCH=AATF" [Accessed February 2011].

146. Online Mendelian Inheritance of Man, Johns Hopkins University. Apoptosis-Antagonizing Transcription Factor; AATF. National Center for Biotechnology Information. "http://www.ncbi.nlm.nih.gov/omim/608463" [Updated February 12, 2004]. [Accessed February 2011].

147. Unigene Organized View Of The Transcriptome. Apoptosis-Associated Tyrosine Kinase [AATK]. National Center for Biotechnology Information. "http://www.ncbi.nlm.nih.gov/UniGene/clust.cgi?UGID=9053 22&TAXID=9606&SEARCH=AATK" [Accessed February 2011].

148. Online Mendelian Inheritance of Man, Johns Hopkins University. Apoptosis-Associated Tyrosine Kinase; AATK. National Center for Biotechnology Information. "http://www.ncbi.nlm.nih.gov/omim/605276" [Updated September 19, 2000]. [Accessed February 2011].

149. Unigene Organized View Of The Transcriptome. Apoptosis Enhancing Nuclease [AEN]. National Center for Biotechnology Information. "http://www.ncbi.nlm.nih.gov/UniGene/clust.cgi?UGID=2311 39&TAXID=9606&SEARCH=AEN" [Accessed February 2011].

150. Online Mendelian Inheritance of Man, Johns Hopkins University. Interferon-Stimulated Exonuclease Gene 20-KD-Like 1; ISG20L1. National Center for Biotechnology Information. "http://www.ncbi.nlm.nih.gov/omim/610177" [Updated June 12, 2006]. [Accessed February 2011].

151. Unigene Organized View Of The Transcriptome. Caspase 1, Apoptosis-Related Cysteine Peptidase [Interleukin 1, Beta, Convertase] [CASP1]. National Center for Biotechnology Information. "http://www.ncbi.nlm.nih.gov/UniGene/clust.cgi?UGID=1312

47&TAXID=9606&SEARCH=Caspase%201"   [Accessed February 2011].

152. Online Mendelian Inheritance of Man, Johns Hopkins University. Caspase 1, Apoptosis Related Cysteine Protease; CASP1. National Center for Biotechnology Information. "http://www.ncbi.nlm.nih.gov/omim/147678"   [Updated January 28, 2011]. [Accessed February 2011].

153. Unigene Organized View Of The Transcriptome. Caspase 2, Apoptosis-Related Cysteine Peptidase [CASP2]. National Center for Biotechnology Information. "http://www.ncbi.nlm.nih.gov/UniGene/clust.cgi?UGID=1978 29&TAXID=9606&SEARCH=CASP2"   [Accessed February 2011].

154. Online Mendelian Inheritance of Man, Johns Hopkins University. Caspase 2, Apoptosis Related Cysteine Protease; CASP2. National Center for Biotechnology Information. "http://www.ncbi.nlm.nih.gov/omim/600639"   [Updated March 3, 2005]. [Accessed February 2011].

155. Unigene Organized View Of The Transcriptome. Caspase 3, Apoptosis-Related Cysteine Peptidase [CASP3]. National Center for Biotechnology Information. "http://www.ncbi.nlm.nih.gov/UniGene/clust.cgi?UGID=1510 37&TAXID=9606&SEARCH=CASP3"   [Accessed February 2011].

156. Online Mendelian Inheritance of Man, Johns Hopkins University. Caspase 3, Apoptosis Related Cysteine Protease; CASP3.   National Center for Biotechnology Information. "http://www.ncbi.nlm.nih.gov/omim/600636"   [Updated December 28, 2010]. [Accessed February 2011].

157. Unigene Organized View Of The Transcriptome. Caspase 4, Apoptosis-Related Cysteine Peptidase [CASP4]. National Center for Biotechnology Information. "http://www.ncbi.nlm.nih.gov/UniGene/clust.cgi?UGID=1508 94&TAXID=9606&SEARCH=CASP4"   [Accessed February 2011].

158. Online Mendelian Inheritance of Man, Johns Hopkins University. Caspase 4, Apoptosis Related Cysteine Protease; CASP4. National Center for Biotechnology Information. "http://www.ncbi.nlm.nih.gov/omim/602664" [Updated May 30, 1998]. [Accessed February 2011].

159. Unigene Organized View Of The Transcriptome. Caspase 5, Apoptosis-Related Cysteine Peptidase [CASP5]. National Center for Biotechnology Information. "http://www.ncbi.nlm.nih.gov/UniGene/clust.cgi?UGID=1647

55&TAXID=9606&SEARCH=CASP5" [Accessed February 2011].

160. Online Mendelian Inheritance of Man, Johns Hopkins University. Caspase 5, Apoptosis Related Cysteine Protease; CASP5. National Center for Biotechnology Information. "http://www.ncbi.nlm.nih.gov/omim/602665" [Updated May 28, 1998]. [Accessed February 2011].

161. Unigene Organized View Of The Transcriptome. Caspase 6, Apoptosis-Related Cysteine Peptidase [CASP6]. National Center for Biotechnology Information. "http://www.ncbi.nlm.nih.gov/UniGene/clust.cgi?UGID=2723 934&TAXID=9606&SEARCH=CASP6" [Accessed February 2011].

162. Online Mendelian Inheritance of Man, Johns Hopkins University. Caspase 6, Apoptosis Related Cysteine Protease; CASP6. National Center for Biotechnology Information. "http://www.ncbi.nlm.nih.gov/omim/601532" [Updated March 10, 2009]. [Accessed February 2011].

163. Unigene Organized View Of The Transcriptome. Caspase 7, Apoptosis-Related Cysteine Peptidase [CASP7]. National Center for Biotechnology Information. "http://www.ncbi.nlm.nih.gov/UniGene/clust.cgi?UGID=1324 72&TAXID=9606&SEARCH=CASP7" [Accessed February 2011].

164. Online Mendelian Inheritance of Man, Johns Hopkins University. Caspase 7, Apoptosis Related Cysteine Protease; CASP7. National Center for Biotechnology Information. "http://www.ncbi.nlm.nih.gov/omim/601761" [Updated April 18, 2006]. [Accessed February 2011].

165. Unigene Organized View Of The Transcriptome. Caspase 8, Apoptosis-Related Cysteine Peptidase [CASP8]. National Center for Biotechnology Information. "http://www.ncbi.nlm.nih.gov/UniGene/clust.cgi?UGID=2068 773&TAXID=9606&SEARCH=CASP8" [Accessed February 2011].

166. Online Mendelian Inheritance of Man, Johns Hopkins University. Caspase 8, Apoptosis Related Cysteine Protease; CASP8. National Center for Biotechnology Information. "http://www.ncbi.nlm.nih.gov/omim/601763" [Updated April 28, 2009]. [Accessed February 2011].

167. Unigene Organized View Of The Transcriptome. Caspase 9, Apoptosis-Related Cysteine Peptidase [CASP9]. National Center for Biotechnology Information. "http://www.ncbi.nlm.nih.gov/UniGene/clust.cgi?UGID=1884

41&TAXID=9606&SEARCH=CASP9" [Accessed February 2011].

168. Online Mendelian Inheritance of Man, Johns Hopkins University. Caspase 9, Apoptosis Related Cysteine Protease; CASP9. National Center for Biotechnology Information. "http://www.ncbi.nlm.nih.gov/omim/602234" [Updated August 2, 2009]. [Accessed February 2011].

169. Unigene Organized View Of The Transcriptome. Caspase 10, Apoptosis-Related Cysteine Peptidase [CASP10]. National Center for Biotechnology Information. "http://www.ncbi.nlm.nih.gov/UniGene/clust.cgi?UGID=1316 69&TAXID=9606&SEARCH=CASP10" [Accessed February 2011].

170. Online Mendelian Inheritance of Man, Johns Hopkins University. Caspase 10, Apoptosis Related Cysteine Protease; CASP10. National Center for Biotechnology Information. "http://www.ncbi.nlm.nih.gov/omim/601762" [Updated October 28, 2008]. [Accessed February 2011].

171. Unigene Organized View Of The Transcriptome. Caspase 14, Apoptosis-Related Cysteine Peptidase [CASP14]. National Center for Biotechnology Information. "http://www.ncbi.nlm.nih.gov/UniGene/clust.cgi?UGID=6812 16&TAXID=9606&SEARCH=CASP14" [Accessed February 2011].

172. Online Mendelian Inheritance of Man, Johns Hopkins University. Caspase 14, Apoptosis Related Cysteine Protease; CASP14. National Center for Biotechnology Information. "http://www.ncbi.nlm.nih.gov/omim/605848" [Updated March 3, 2008]. [Accessed February 2011].

173. Unigene Organized View Of The Transcriptome. Dysbindin (Dystrobrevin Binding Protein 1) Domain Containing 2 [DBNDD2]. National Center for Biotechnology Information. "http://www.ncbi.nlm.nih.gov/UniGene/clust.cgi?UGID=2724 373&TAXID=9606&SEARCH=DBNDD2" [Accessed February 2011].

174. Online Mendelian Inheritance of Man, Johns Hopkins University. Dysbindin Domain-Containing 2; DBNDD2. National Center for Biotechnology Information. "http://www.ncbi.nlm.nih.gov/omim/611453" [Updated September 20, 2007]. [Accessed February 2011].

175. Unigene Organized View Of The Transcriptome. Fas [TNFRSF6]-Associated Via Death Domain [FADD]. National Center for Biotechnology Information. "http://www.ncbi.nlm.nih.gov/UniGene/clust.cgi?UGID=1409 19&TAXID=9606&SEARCH=FADD" [Accessed February 2011].

176. Online Mendelian Inheritance of Man, Johns Hopkins University. FAS-Associated Via Death Domain; FADD. National Center for Biotechnology Information. "http://www.ncbi.nlm.nih.gov/omim/602457" [Updated November 30, 2009]. [Accessed February 2011].

177. Unigene Organized View Of The Transcriptome. Leucine-Rich Repeats and Death Domain Containing [LRDD]. National Center for Biotechnology Information. "http://www.ncbi.nlm.nih.gov/UniGene/clust.cgi?UGID=2061 301&TAXID=9606&SEARCH=PIDD" [Accessed February 2011].

178. Online Mendelian Inheritance of Man, Johns Hopkins University. Leucine-Rich Repeats and Death Domain Containing; LRDD. National Center for Biotechnology Information. "http://www.ncbi.nlm.nih.gov/omim/605247" [Updated May 12, 2005]. [Accessed February 2011].

179. Unigene Organized View Of The Transcriptome. THAP Domain Containing, Apoptosis Associate Protein 1 [THAP1]. National Center for Biotechnology Information. "http://www.ncbi.nlm.nih.gov/UniGene/clust.cgi?UGID=1321 26&TAXID=9606&SEARCH=THAP1" [Accessed February 2011].

180. Online Mendelian Inheritance of Man, Johns Hopkins University. THAP Domain-Containing Protein 1; THAP1. National Center for Biotechnology Information. "http://www.ncbi.nlm.nih.gov/omim/609520" [Updated June 25, 2010]. [Accessed February 2011].

181. Unigene Organized View Of The Transcriptome. Hypothetical Protein LOC100289049 [LOC100289049]. National Center for Biotechnology Information. "http://www.ncbi.nlm.nih.gov/UniGene/clust.cgi?UGID=1686 92&TAXID=9606&SEARCH=THAP2" [Accessed February 2011].

182. Online Mendelian Inheritance of Man, Johns Hopkins University. THAP Domain-Containing Protein 2; THAP2. National Center for Biotechnology Information. "http://www.ncbi.nlm.nih.gov/omim/612531" [Updated January 16, 2009]. [Accessed February 2011].

183. Unigene Organized View Of The Transcriptome. THAP Domain Containing, Apoptosis Associate Protein 3 [THAP3]. National Center for Biotechnology Information. "http://www.ncbi.nlm.nih.gov/UniGene/clust.cgi?UGID=3321 862&TAXID=9606&SEARCH=THAP3" [Accessed July 2009].

184. Online Mendelian Inheritance of Man, Johns Hopkins University. THAP Domain-Containing Protein 3; THAP3.

National Center for Biotechnology Information. "http://www.ncbi.nlm.nih.gov/omim/612532" [Updated January 16, 2009]. [Accessed February 2011].

185. Unigene Organized View Of The Transcriptome. TNFRSF1A-Associated Via Death Domain [TRADD]. National Center for Biotechnology Information. "http://www.ncbi.nlm.nih.gov/UniGene/clust.cgi?UGID=6761 55&TAXID=9606&SEARCH=TRADD" [Accessed February 2011].

186. Online Mendelian Inheritance of Man, Johns Hopkins University. Tumor Necrosis Factor Receptor 1-Associated Death Domain Protein; TRADD. National Center for Biotechnology Information. "http://www.ncbi.nlm.nih.gov/omim/603500" [Updated April 29, 2004]. [Accessed February 2011].

187. Cancer GeneticsWeb. Bladder Cancer. Cancer GeneticsWeb. "http://www.cancerindex.org/geneweb/X2103.htm" [Accessed February 2011].

188. Cancer GeneticsWeb. Osteosarcoma. Cancer GeneticsWeb. "http://www.cancerindex.org/geneweb/X0203.htm" [Accessed February 2011].

189. Cancer GeneticsWeb. Brain and CNS Tumours. Cancer GeneticsWeb. "http://www.cancerindex.org/geneweb/X0301.htm" [Accessed February 2011].

190. Cancer GeneticsWeb. Breast Cancer. Cancer GeneticsWeb. "http://www.cancerindex.org/geneweb/X0401.htm" [Accessed February 2011].

191. Cancer GeneticsWeb. ColoRectal Cancers. Cancer GeneticsWeb. "http://www.cancerindex.org/geneweb/X0501.htm" [Accessed February 2011].

192. Cancer GeneticsWeb. Renal Cell Carcinoma. Cancer GeneticsWeb. "http://www.cancerindex.org/geneweb/X210201.htm" [Accessed February 2011].

193. Cancer GeneticsWeb. Acute Lymphocytic Leukaemia. Cancer GeneticsWeb. "http://www.cancerindex.org/geneweb/X1205.htm" [Accessed February 2011].

194. Cancer GeneticsWeb. Acute Myeloid Leukaemia. Cancer GeneticsWeb. "http://www.cancerindex.org/geneweb/X1206.htm" [Accessed February 2011].

195. Cancer GeneticsWeb. Chronic Lymphocytic Leukaemia. Cancer GeneticsWeb. "http://www.cancerindex.org/geneweb/X1207.htm" [Accessed February 2011].

196. Cancer GeneticsWeb. Chronic Myeloid Leukaemia. Cancer GeneticsWeb. "http://www.cancerindex.org/geneweb/X1208.htm" [Accessed February 2011].

197. Cancer GeneticsWeb. Hairy Cell Leukaemia. Cancer GeneticsWeb. "http://www.cancerindex.org/geneweb/X1209.htm" [Accessed February 2011].

198. Cancer GeneticsWeb. Hepatocelular Carcinoma. Cancer GeneticsWeb. "http://www.cancerindex.org/geneweb/X070601.htm" [Accessed February 2011].

199. Cancer GeneticsWeb. Lung Cancer. Cancer GeneticsWeb. "http://www.cancerindex.org/geneweb/X1501.htm" [Accessed February 2011].

200. Cancer GeneticsWeb. Hodgkin's Disease. Cancer GeneticsWeb. "http://www.cancerindex.org/geneweb/X1602.htm" [Accessed February 2011].

201. Cancer GeneticsWeb. Non-Hodgkin's Lymphoma. Cancer GeneticsWeb. "http://www.cancerindex.org/geneweb/X1603.htm" [Accessed February 2011].

202. Cancer GeneticsWeb. Multiple Myeloma. Cancer GeneticsWeb. "http://www.cancerindex.org/geneweb/X1301.htm" [Accessed February 2011].

203. Cancer GeneticsWeb. Ovarian Cancer. Cancer GeneticsWeb. "http://www.cancerindex.org/geneweb/X1003.htm" [Accessed February 2011].

204. Cancer GeneticsWeb. Pancreatic Cancer. Cancer GeneticsWeb. "http://www.cancerindex.org/geneweb/X0603.htm" [Accessed February 2011].

205. Cancer GeneticsWeb. Prostate Cancer. Cancer GeneticsWeb. "http://www.cancerindex.org/geneweb/X0904.htm" [Accessed February 2011].

206. Cancer GeneticsWeb. Melanoma. Cancer GeneticsWeb. "http://www.cancerindex.org/geneweb/X1902.htm" [Accessed February 2011].

207. Cancer GeneticsWeb. Non-Melanoma Skin Cancer. Cancer GeneticsWeb. "http://www.cancerindex.org/geneweb/X1903.htm" [Accessed February 2011].

208. Cancer GeneticsWeb. Stomach (Gastric) Cancer. Cancer GeneticsWeb. "http://www.cancerindex.org/geneweb/X0702.htm" [Accessed February 2011].

209. Cancer GeneticsWeb. Testicular Cancer. Cancer GeneticsWeb. "http://www.cancerindex.org/geneweb/X0902.htm" [Accessed February 2011].

210. Cancer GeneticsWeb. Thyroid Cancer. Cancer GeneticsWeb. "http://www.cancerindex.org/geneweb/X0605.htm" [Accessed February 2011].

211. Cancer GeneticsWeb. Uterine Cancer. Cancer GeneticsWeb. "http://www.cancerindex.org/geneweb/X1005.htm" [Accessed February 2011].

212. Tong Q-S, Zheng L-D, Wang L, Liu J, Qian W. BAK overexpression mediates p53-independent apoptosis inducing effects on human gastric cancer cells. "BMC Cancer". 2004; 4:33.

213. Bougras G, Cartron P-F, Gautier F, et al. Opposite role of Bax and BCL-2 in the anti-tumoral responses of the immune system." BMC Cancer". 2004; 4:54.

214. Hao H, Dong Y, Bowling MT, Gomez-Gutierrez JG, Zhou HS, McMasters KM. E2F-1 induces melanoma cell apoptosis via PUMA up-regulation and Bax translocation. "BMC Cancer". 2007; 7:24.

215. Wei MC, Zong W-X, Cheng E H-Y, et al. Proapoptotic BAX and BAK: A Requisite Gateway to Mitochondrial Dysfunction and Death. *Science*. 2001; 292:727-730.

216. Zong W-X, Li C, Hatzivassiliou G, et al. Bax and Bak can localize to the endoplasmic reticulum to initiate apoptosis. "The Journal of Cell Biology". 2003; 162:59-69.

217. Coffin JM, Hughes SH, Varmus HE, eds. "Retroviruses at The National Center for Biotechnology Information" [book online]. Cold Spring Harbor, NY: Cold Spring Harbor Laboratory Press; 1997. "http://www.ncbi.nlm.nih.gov/bookshelf/br.fcgi?book=rv&part=A288" [Accessed February 2011].

218. MicrobiologyBytes. Retroviruses. MicrobiologyBytes. "http://www.microbiologybytes.com/virology/Retroviruses.html" [Updated April 8, 2009]. [Accessed February 2011].

219. Schmidt, Elaine. UCLA Scientists Transform HIV Into Cancer-Seeking Missile; Firefly Protein Illuminates Virus' Hunt of

Metastasized Melanoma Cells in Live Mouse. "UCLA Newsroom".
"http://newsroom.ucla.edu/page.asp?RelNum=5921"
[Published February 14, 2005].

220. UniprotKB. Gag-Pro-Pol Polyprotein – Human T-Cell Leukemia Virus 2 [HTLV-2]. UniprotKB. "http://www.uniprot.org/uniprot/P03363" [Updated February 8, 2011]. [Accessed February 2011].

221. UniprotKB. Envelope Glycoprotein gp63 Precursor – Human T-Cell Leukemia Virus 2 [HTLV-2]. UniprotKB. "http://www.uniprot.org/uniprot/P03383" [Updated February 8, 2011]. [Accessed February 2011].

222. UniprotKB. Protein Tax-2 – Human T-Cell Leukemia Virus 2 [HTLV-2]. UniprotKB. "http://www.uniprot.org/uniprot/P03410" [Updated February 8, 2011]. [Accessed February 2011].

223. UniprotKB. Protein Rex – Human T-Cell Leukemia Virus 2 [HTLV-2]. UniprotKB. "http://www.uniprot.org/uniprot/Q85601" [Updated November 30, 2010]. [Accessed February 2011].

Picture References

1p. Research Apoptosis. Intrinsic Apoptosis Diagram. Research Apoptosis. "http://www.researchapoptosis.com/apoptosis/pathways/intrinsic/index.m" [Published 2002]. [Accessed February 2011].

2p. Research Apoptosis. Extrinsic Apoptosis Diagram. Research Apoptosis. "http://www.researchapoptosis.com/apoptosis/pathways/extrinsic/index.m" [Published 2002]. [Accessed February 2011].

3p. Access Excellence at the National Health Museum Resource Center. Retrovirus Replication. Access Excellence at the National Health Museum Resource Center. "http://www.accessexcellence.org/RC/VL/GG/retrovirus.php" [Accessed February 2011].

4p. Gene Therapy Review. Retrovirus Genome Diagram. Gene Therapy Review. "http://www.genetherapyreview.com/gene-therapy-education/gene-transfer-vectors/1/8-retrovirus.html" [Published September 5, 2008]. [Accessed February 2011].

5p. MicrobiologyBytes. Retroviruses. MicrobiologyBytes. "http://www.microbiologybytes.com/virology/Retroviruses.html" [Updated April 8, 2009]. [Accessed February 2011].

6p. Viral Zone. Human T-Cell Leukemia Virus 2, Molecular Biology Virion and Genome. Viral Zone. "http://www.expasy.ch/viralzone/all_by_protein/61.html" [Accessed February 2011].